SPACEPOSIUM

The New Age of Space Technology

By
Ben McLaren

First Edition
2025

SPACEPOSIUM
THE NEW AGE OF SPACE TECHNOLOGY

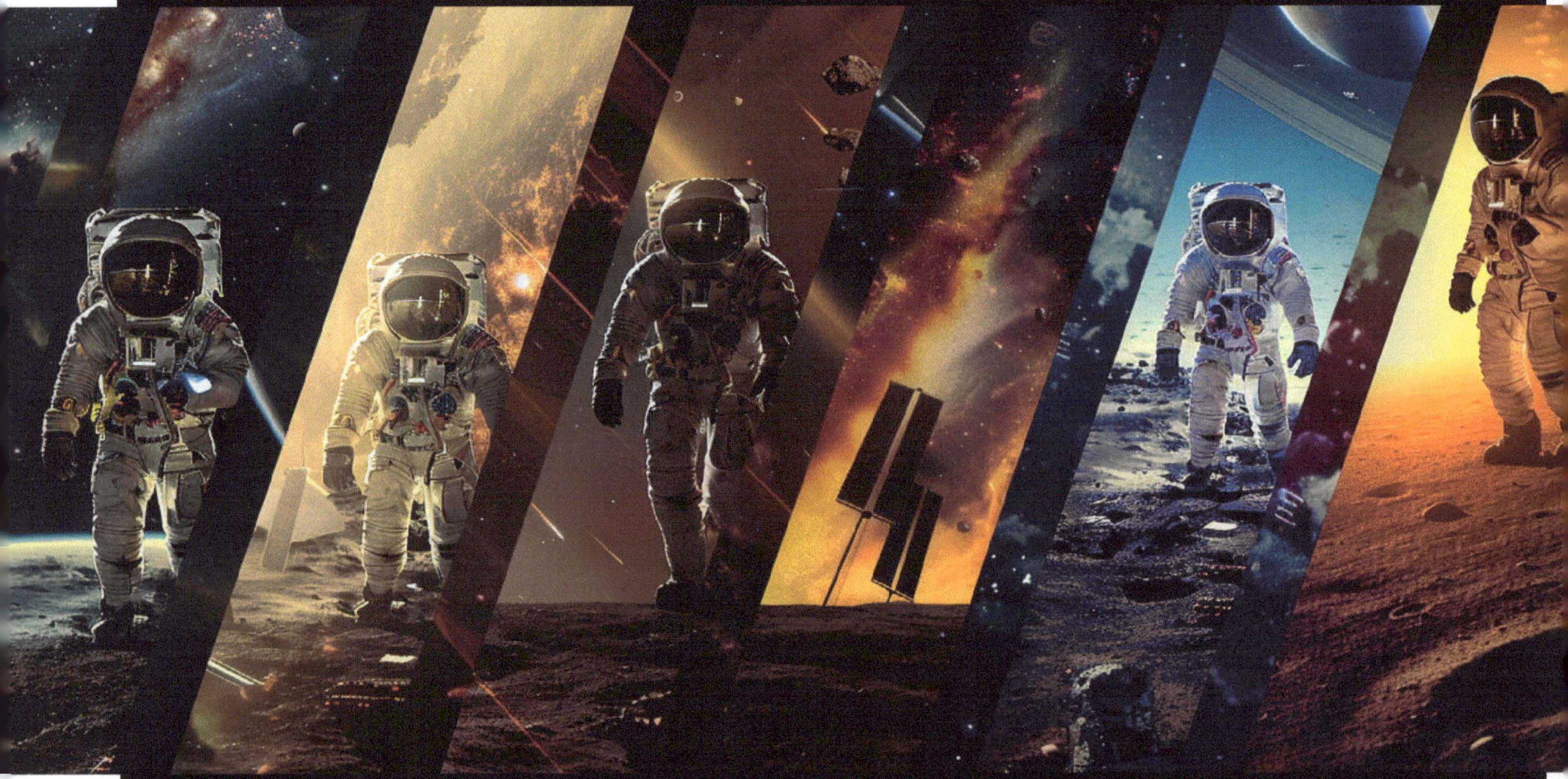

Ben McLaren
SPACEPOSIUM

ISBN 979-8-89691-132-6

SPACEPOSIUM
THE NEW AGE OF SPACE TECHNOLOGY

SPACEPOSIUM
TABLE OF CONTENTS

SPACEPOSIUM

SPACEPOSIUM

FORWARD

Welcome to the dawn of a new era in human history. As we stand on the cusp of unprecedented exploration and development beyond our home planet, Spaceposium arrives at just the right moment to shine a light on the challenges, opportunities, and profound implications of our expanding cosmic frontier.

The 21st century has ignited a renaissance in space activities. Visionary private companies are joining forces with national space agencies, pushing the boundaries of what we once thought possible. Imagine standing on the surface of Mars, gazing at a sky filled with unfamiliar stars, or witnessing the Earth rise from the Moon's horizon. Within our lifetimes, humanity is poised to become a truly multi-planetary species.

But this thrilling new Space Age also presents complex questions and formidable hurdles. How can we sustainably harness space resources? What legal and ethical frameworks will govern our actions in the cosmos? How do we ensure that space remains accessible and equitable for all of humanity? Spaceposium dives headfirst into these critical issues, offering a comprehensive and thought-provoking examination of our spacefaring future.

This book unites a diverse array of perspectives from experts in fields—ranging from aerospace engineering and planetary science to space law and futurism. It provides readers with a sweeping view of humanity's next great leap. From the technical intricacies of long-duration spaceflight to the societal impacts of establishing off-world settlements, Spaceposium leaves no stone unturned in its exploration of our cosmic destiny.

Spaceposium is an invaluable resource that crystallizes the key debates and developments that will shape our future in space. Whether you are a seasoned space professional, an aspiring astronaut, or a curious earthling, this book will expand your understanding of the final frontier and humanity's place within it.

The journey to the stars promises to be the greatest adventure in human history. Spaceposium is your guide to understanding the monumental voyage that lies ahead. As we embark on this incredible journey, this book illuminates the path forward into the vast, uncharted cosmic ocean.

The future of space exploration is here, and it's more exciting than we ever imagined. Together, we can achieve what was once thought impossible and leave a lasting legacy for future generations. The universe awaits—let's explore.

Our spacecraft are more than machines...
they are dreams welded into metal and hope.

The universe does not invite us.
It challenges us to understand.

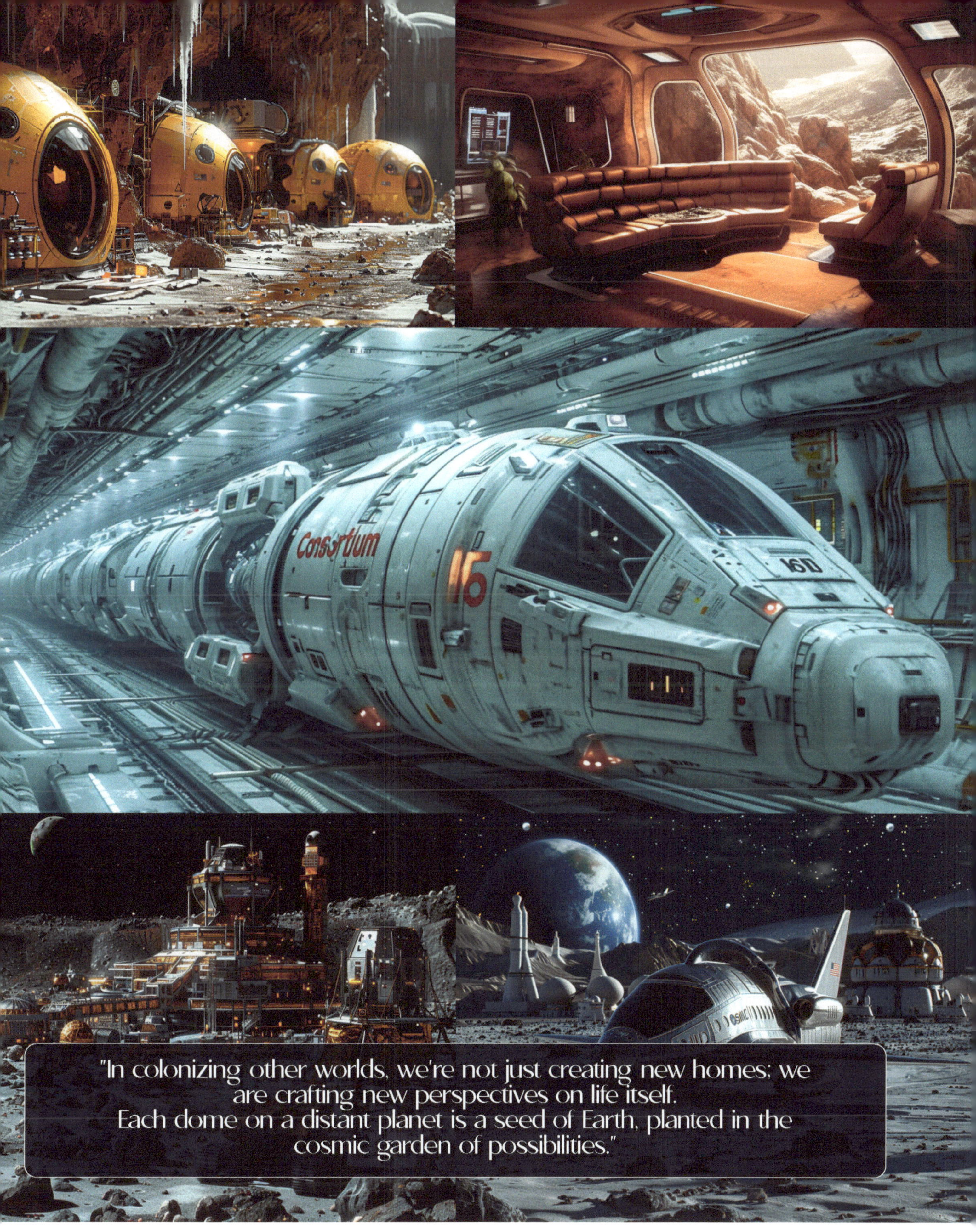

"In colonizing other worlds, we're not just creating new homes; we are crafting new perspectives on life itself.
Each dome on a distant planet is a seed of Earth, planted in the cosmic garden of possibilities."

SPACEPOSIUM

CHAPTER 1 FOUNDATIONS OF SPACE EXPLORATION

"The frontier of space is where we rewrite the boundaries of what's possible, one mission at a time."

THE NEW FRONTIER—AN OVERVIEW OF TODAY'S SPACE INDUSTRY

What lies beyond the boundaries of our planet? This question has captivated humanity since we first gazed at the stars. Today, we stand on the cusp of a new era in space exploration, one that promises to redefine our relationship with the cosmos.

From ancient civilizations charting the heavens to the dawn of modern rocketry, from the Cold War space race to the current age of commercial spaceflight, our journey to the stars has been a testament to human ingenuity and perseverance.

Get ready to be propelled into an extraordinary voyage through time, tracing humanity's relentless pursuit of the cosmos from our earliest imaginings to the moment we first broke free from Earth's gravitational embrace.

In just a few decades, we've progressed from launching the first artificial satellite to planning missions to Mars.

Private companies now compete alongside nations in the space race, driving innovation at an unprecedented pace. The James Webb Space Telescope peers deeper into the universe than ever before, while reusable rockets make space more accessible than we ever imagined.

This book, "Spaceposium: The New Age of Space Technology," takes you on a journey through the cutting-edge advancements in technology that are now shaping our future.

Starting with a brief look back at the history of ancient astronomical knowledge that laid the foundation for modern space exploration. We are going to explore these areas and more:

- Ancient astronomers who built and inhabited megalithic sites.
- Visionaries who transformed space travel from fantasy to reality.
- The role of rocketry in enabling our journey beyond Earth.
- The Space Race and its lasting impact on space technology.
- The current Space Industry and the rise of commercial spaceflight.
- Scientific discoveries and technological innovations of space exploration.
- The economic benefits and spin-off technologies that improve life on Earth.
- Cultural impacts of space exploration, inspiring unity and influencing media.

As we embark on this exploration, we'll see how the integration of space technology is revolutionizing life on Earth, influencing everything from agriculture to education. We'll also look ahead to the future, envisioning how continued advancements will drive further exploration and create new opportunities for humanity.

Since the dawn of consciousness, humans have gazed at the night sky with wonder and longing. The vast cosmic tapestry above us has inspired dreams, myths, and an unquenchable desire to reach beyond our earthly bonds. We embark on an extraordinary journey through time, tracing humanity's relentless pursuit of the stars from our earliest imaginings to the moment we first broke free from Earth's gravitational embrace.

THE COSMIC DREAM: FROM MYTH TO REALITY

For millennia, the heavens have beckoned to us, sparking tales of celestial gods, heroic ascents, and worlds beyond our own. From the winged sandals of Hermes to the sky-chariot of Indra, every culture on Earth has woven stories of transcending our terrestrial limits. These myths were more than mere fancy – they were the first embers of a fire that would eventually propel us into the cosmos itself.
As we explore these ancient dreams, we'll see how they laid the psychological groundwork for humanity's greatest adventure.

THE GLOBAL QUEST FOR CELESTIAL KNOWLEDGE

Long before we could touch the stars, we strove to understand them. This insatiable curiosity gave birth to astronomy – perhaps humanity's oldest science. From Mayan observatories to Stonehenge, from ancient Babylonian clay tablets to the intricate astrolabes of medieval Islam, we'll discover how civilizations across the globe and throughout history painstakingly charted the heavens. These early astronomers, separated by vast distances but united in purpose, created the foundation of knowledge that would one day allow us to navigate the ocean of space.

As we delve into these ancient achievements, we'll see how they converged in the 20th century, igniting a technological revolution that would finally make our oldest dreams a reality. The story of space exploration is the story of humanity itself—our boundless curiosity, remarkable ingenuity, and unyielding spirit of adventure. Let's explore the past for a moment.

GLOBAL CONTRIBUTIONS TO ANCIENT ASTRONOMY

1. Mesopotamia: Often regarded as the birthplace of astronomy, Mesopotamian civilizations such as the Sumerians, Babylonians, and Assyrians made early advancements in tracking celestial bodies. They developed the zodiac, recorded planetary movements, and created mathematical models to predict eclipses and other astronomical events. Their sexagesimal (base-60) number system is still used today in measuring time and angles.

2. Egypt: Ancient Egyptian astronomy was closely tied to their agricultural calendar and religious practices. They developed a 365-day calendar based on the heliacal rising of Sirius and aligned their monumental structures, such as the pyramids, with cardinal points and significant stars. Egyptian astronomers also made early observations of variable stars like Algol, demonstrating their sophisticated understanding of celestial phenomena.

3. China: Chinese astronomers meticulously documented celestial events such as eclipses, comets, and supernovae. They developed a lunisolar calendar and were among the first to record sunspots and supernovae. Chinese star maps and catalogs, such as those created by Gan De and Shi Shen, reflect their advanced observational techniques and their influence on later astronomical practices.

4. India: Indian astronomers made significant contributions through texts like the Vedanga Jyotisha and the works of scholars such as Aryabhata and Brahmagupta. They proposed early heliocentric models, calculated the Earth's rotation period, and developed sophisticated mathematical techniques for predicting celestial movements. These contributions were crucial in the development of both astronomy and mathematics.

5. Greece: Greek astronomers, building on knowledge from Babylonian and Egyptian sources, made foundational contributions to Western astronomy. Figures like Ptolemy, Hipparchus, and Aristarchus developed geometric models of the cosmos, cataloged stars, and proposed early heliocentric theories. The Greek tradition of inquiry and synthesis significantly influenced Islamic and later European astronomy.

6. Mesoamerica: The Maya and Aztec civilizations developed complex calendar systems and built observatories to track celestial events. The Maya, in particular, made precise observations of Venus and solar eclipses, which were integral to their calendar and agricultural practices. Their astronomical knowledge was deeply intertwined with their cultural and religious life.

7. Islamic Golden Age: Islamic astronomers preserved and expanded upon Greek and Indian astronomical knowledge. Scholars like Al-Battani, Ibn al-Haytham, and Nasir al-Din al-Tusi made significant advancements in observational techniques, mathematical models, and the development of astronomical instruments. Their work laid the groundwork for the Renaissance and the eventual transition to modern astronomy.

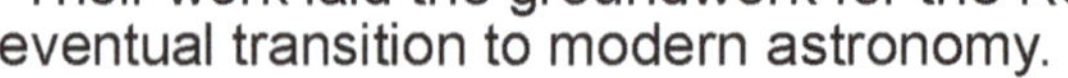

THE TIME LINES OF ANCIENT ASTRONOMY

Throughout time, cultures came and went, but many contributed unique perspectives and significant discoveries, from the mathematical precision of Babylonian astronomy to the observational rigor of Chinese and Islamic scholars. Let's look at it from a time-line perspective:

Global Milestones of Ancient Astronomy:

1. Prehistoric Astronomy (11,000 BCE - 3000 BCE)
- Nabta Playa, Egypt (c. 11,000 BCE): Oldest known astronomical site, featuring stone circles and calendar system.
- Stonehenge, England (c. 3000 BCE): Megalithic structure aligned with solstices and equinoxes.

2. Mesopotamian Astronomy (3000 BCE - 100 CE)
- Sumerians (c. 3000 BCE): Developed early concepts of constellations and planetary movements.
- Babylonians (c. 1800 BCE): Created sophisticated mathematical models for predicting celestial events.
- Key contributions: Sexagesimal number system, **zodiac**, and eclipse predictions.

3. Egyptian Astronomy (3000 BCE - 30 BCE)
- Aligned pyramids with cardinal directions and stars (c. 2600 BCE).
- Developed 365-day calendar based on Sirius's heliacal rising.
- Divided the night sky into 36 groups of stars (decans) for timekeeping.

4. Chinese Astronomy (2000 BCE - present)
- Recorded solar eclipses as early as 2137 BCE.
- Developed accurate calendar systems and cataloged supernovae.
- Invented seismoscopes (132 CE) and built sophisticated observatories.

5. Indian Astronomy (1500 BCE - 1200 CE)
- Vedanga Jyotisha (c. 1400 BCE): Early astronomical text detailing calendars and celestial movements.
- Aryabhata (476-550 CE): Proposed heliocentric model and calculated Earth's rotation period.
- Brahmagupta (598-668 CE): Introduced mathematical astronomy and gravity concepts.

6. Mesoamerican Astronomy (1500 BCE - 1500 CE)
- **Maya civilization (c. 2000 BCE - 1500 CE):**
  - Developed complex calendar systems (Tzolk'in and Haab').
  - Built observatories like El Caracol at Chichen Itza.
  - Accurately predicted Venus's movements and solar eclipses.
- **Aztec civilization (14th-16th century CE):**
  - Created Sun Stone calendar.
  - Observed and recorded celestial events.

7. Greek Astronomy (600 BCE - 400 CE)
- Thales of Miletus (c. 624-546 BCE): Predicted solar eclipses.
- Aristotle (384-322 BCE): Proposed geocentric model of the universe.
- Hipparchus (190-120 BCE): Created star catalog and discovered procession of equinoxes.
- Ptolemy (100-170 CE): Compiled Almagest, the most influential ancient astronomical text.

8. South American Astronomy (400 BC-1532 CE)
- Nazca Lines, Peru (400 BCE - 650 CE):
- Inca civilization (1400 1532 CE):
  - Built astronomical observatories like Machu Picchu.

9. Roman Astronomy (753 BCE - 476 CE)
- Adopted and adapted Greek astronomical knowledge.
- Julius Caesar (100-44 BCE): Introduced Julian calendar.

10. Islamic Golden Age Astronomy (8th-14th century CE)
- Al-Battani (858-929): Improved astronomical calculations and instruments.
- Ibn al-Haytham (965-1040): Made significant contributions to optics, scientific method.

11. Renaissance Astronomy (14th-17th century)
- Nicolaus Copernicus (1473-1543): Proposed heliocentric model of the solar system.
- Galileo Galilei (1564-1642): Pioneered use of the telescope for astronomical observations, supported heliocentrism.
- Johannes Kepler (1571-1630): Formulated laws of planetary motion.

THE ADVENT OF ROCKETS

The role of rocketry in enabling space exploration cannot be overstated. Rockets provided the means to overcome Earth's gravitational pull and propel objects into space. The history of rocketry dates back to ancient China, where gunpowder-powered rockets were used for military purposes as early as the 13th century. However, it wasn't until the 20th century that rockets were developed specifically for space exploration.

Early myths about space have been part of human culture for millennia. Greek mythology tells the tale of Daedalus and Icarus, who attempted to fly using wings made of feathers and wax. This story reflects humanity's ancient desire to break free from Earth's gravity and soar into the heavens. As science and technology progressed, these dreams evolved from myth to speculation, with writers like Jules Verne (1828-1905) and H.G. Wells (1866-1946) imagining fantastic voyages beyond our planet.

Key figures in early rocketry played pivotal roles in transforming space travel from fantasy to reality:

1. Konstantin Tsiolkovsky (1857-1935): Often called the father of spaceflight, this Russian school teacher proposed the idea of space exploration by rocket in 1898. Tsiolkovsky developed the fundamental equations of rocket propulsion and envisioned multi-stage rockets, space stations, and the use of liquid propellants.

2. Robert Goddard (1882-1945): An American engineer and physicist, Goddard is credited with creating and launching the world's first liquid-fueled rocket on March 16, 1926. His work dramatically improved rocket efficiency and laid the groundwork for modern rocketry. Goddard's 1919 monograph "A Method of Reaching Extreme Altitudes" is considered a classic text in 20th-century rocket science.

3. Wernher von Braun (1912-1977): A German-American aerospace engineer, von Braun played a crucial role in the development of rocket technology during World War II and later in the United States space program. He led the design of the Saturn V rocket that took humans to the Moon and was instrumental in popularizing the idea of space exploration.

These pioneers, along with many others, transformed rocketry from a curiosity into the cornerstone of space exploration. Their work paved the way for the Space Race of the 1960s, which saw rapid advancements in rocket technology and culminated in the first human steps on the Moon.

As we delve deeper into the history and future of space exploration, we'll see how the dreams of these early visionaries have shaped our understanding of the cosmos and continue to inspire new generations of scientists, engineers, and explorers to push the boundaries of human knowledge and achievement.

Several key technological advancements enabled the first successful rockets:

1. Development of gunpowder: The invention of gunpowder in ancient China around the 9th century CE laid the foundation for early rocket propulsion.

2. Solid propellants: Early rockets used solid propellants like gunpowder packed into tubes, as seen in Chinese fire arrows and later European military rockets.

3. Metal casings: The transition from bamboo tubes to metal casings in Europe during the Renaissance improved rocket durability and performance.

4. Liquid propellants: Robert Goddard's development of liquid-fueled rockets in the 1920s was a major breakthrough, allowing for greater power and control.

5. De Laval nozzle: Goddard's application of a supersonic (de Laval) nozzle to the combustion chamber significantly improved rocket efficiency.

6. Gyroscopic stabilization: Goddard and others developed gyro-stabilization systems to improve rocket guidance and stability.

7. Multi-stage rockets: The concept of multi-stage rockets, theorized by Konstantin Tsiolkovsky and others, allowed for greater range and payload capacity.

8. Improved fuels: The use of more energetic propellants, such as liquid oxygen and hydrogen, increased rocket performance.

9. Guidance systems: The development of guidance technology, including gyroscopes and later electronic systems, improved accuracy.

10. Aerodynamics: Advancements in understanding aerodynamics led to improved rocket designs, reducing drag and increasing efficiency.

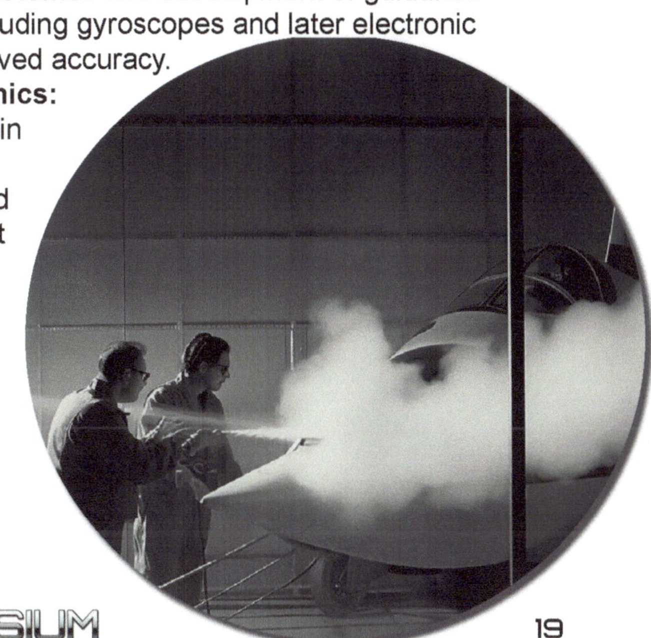

These technological advancements, combined with theoretical work by pioneers like Tsiolkovsky, Goddard, and von Braun, paved the way for the development of modern rocketry and space exploration.

THE SPACE RACE

The Space Race was a pivotal chapter in the history of space exploration, driven by the intense rivalry between the United States and the Soviet Union during the Cold War. This section delves into the Cold War context and its impact on space exploration, the major milestones achieved during this period, and the culmination of the race with the Apollo program and the Moon landings.

Cold War Context and Its Impact on Space Exploration
The Space Race emerged as a critical front in the broader Cold War, a period of geopolitical tension between the United States and the Soviet Union that lasted from the end of World War II until the early 1990s. This ideological battle between capitalism and communism extended into various domains, including military, economic, and technological arenas. Space became a new battleground where each superpower sought to demonstrate its superiority.

The launch of Sputnik 1 by the Soviet Union on October 4, 1957, marked the beginning of the Space Race. This first artificial satellite to orbit Earth sent shockwaves through the United States, igniting fears of Soviet technological and military dominance[1][2]. The successful launch of Sputnik not only showcased Soviet prowess in rocketry but also implied potential advancements in intercontinental ballistic missile (ICBM) technology, heightening Cold War tensions.

In response, the United States accelerated its own space program, leading to the establishment of the National Aeronautics and Space Administration (NASA) in 1958. The competition between the two superpowers spurred rapid advancements in space technology and exploration, with each side striving to achieve significant milestones to assert their dominance on the global stage.

Major Milestones: Sputnik, Yuri Gagarin's Flight, Mercury and Gemini Programs
Sputnik and Early Soviet Achievements

The Soviet Union's early achievements in space were remarkable. Following Sputnik 1, they launched Sputnik 2 on November 3, 1957, carrying the first living creature into space, a dog named Laika. These early successes demonstrated the Soviet Union's lead in space technology and intensified the pressure on the United States to catch up.

Yuri Gagarin's Historic Flight

On April 12, 1961, Soviet cosmonaut Yuri Gagarin became the first human to journey into outer space aboard Vostok 1. Gagarin's flight, which completed one orbit of Earth in 108 minutes, was a monumental achievement that further solidified the Soviet Union's lead in the Space Race. Gagarin's successful mission made him an international hero and underscored the Soviet Union's technological capabilities.

SPACEPOSIUM

Mercury and Gemini Programs

In response to Soviet advancements, the United States launched Project Mercury, aimed at putting an American astronaut into space. On May 5, 1961, Alan Shepard became the first American in space with a suborbital flight aboard Freedom 7. This was followed by John Glenn's orbital flight on February 20, 1962, making him the first American to orbit Earth.

The Gemini program, which ran from 1962 to 1966, served as a crucial bridge between Mercury and the Apollo programs. Gemini missions tested essential technologies and techniques needed for lunar exploration, including long-duration space flights, extravehicular activities (spacewalks), and orbital rendezvous and docking[9][10]. These missions provided invaluable experience and data, paving the way for the Apollo program.

The Apollo Program and the Moon Landings

The Apollo program was the United States' ambitious effort to land humans on the Moon and return them safely to Earth. Announced by President John F. Kennedy in 1961, the program aimed to achieve this goal before the end of the decade. The Apollo missions represented the pinnacle of human space exploration and technological achievement.

Apollo 11: The First Moon Landing

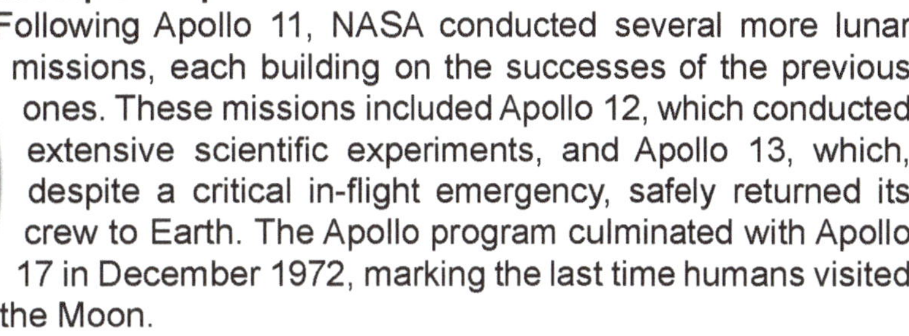

On July 20, 1969, Apollo 11 successfully landed on the Moon, with astronauts Neil Armstrong and Edwin "Buzz" Aldrin becoming the first humans to set foot on the lunar surface. Armstrong's famous words, "That's one small step for man, one giant leap for mankind," echoed around the world, symbolizing a triumph not just for the United States but for all of humanity.

Subsequent Apollo Missions

Following Apollo 11, NASA conducted several more lunar missions, each building on the successes of the previous ones. These missions included Apollo 12, which conducted extensive scientific experiments, and Apollo 13, which, despite a critical in-flight emergency, safely returned its crew to Earth. The Apollo program culminated with Apollo 17 in December 1972, marking the last time humans visited the Moon.

The achievements of the Apollo program demonstrated the United States' technological and scientific capabilities, effectively ending the Space Race in favor of the United States. The legacy of the Apollo missions continues to inspire future generations of space explorers and scientists.

Post-Apollo Era

The post-Apollo era saw NASA shift its focus towards reusable spacecraft, international cooperation in space, and robotic exploration of the solar system. This period marked a significant transition in space exploration, with increased involvement from private companies and international partners.

Space Shuttle Program

The Space Shuttle program, which ran from 1981 to 2011, introduced the world's first reusable spacecraft. Key features of the program included:

- Reusability: The orbiter could be used for multiple missions, reducing costs compared to expendable rockets.
- Versatility: It served as a satellite launch vehicle, research platform, and cargo transport to space stations.
- Payload capacity: The Shuttle could carry up to 24,400 kg to low Earth orbit.
- Scientific contributions: It deployed and serviced the Hubble Space Telescope and conducted numerous microgravity experiments.
- Space station support: The Shuttle played a crucial role in assembling and supplying the International Space Station.

Major companies involved in the Space Shuttle program included:
1. Boeing (www.boeing.com/space): Prime contractor for the Space Shuttle, responsible for its manufacture, integration, and operation.
2. Lockheed Martin (www.lockheedmartin.com/en-us/capabilities/space.html): Developed the external tank and supported various shuttle operations.
3. Northrop Grumman (www.northropgrumman.com/space): Manufactured the solid rocket boosters and provided engineering support.

International Cooperation: Mir and the International Space Station

The post-Apollo era saw increased international collaboration in space exploration:
Mir Space Station:
- Operated by the Soviet Union (later Russia) from 1986 to 2001
- Hosted international crews, including American astronauts
- Served as a precursor to the International Space Station

International Space Station (ISS):
- A joint project involving the United States, Russia, Europe, Japan, and Canada
- Assembly began in 1998 and has been continuously occupied since 2000
- Serves as a microgravity and space environment research laboratory

Key companies involved in the ISS program include:
1. Boeing (www.boeing.com/space/international-space-station): Prime contractor for the U.S. segment of the ISS.
2. Thales Alenia Space (www.thalesgroup.com/en/space): Manufactured several ISS modules for NASA and ESA.
3. Roscosmos State Corporation (www.roscosmos.ru): Responsible for the Russian segment of the ISS.
4. SpaceX (www.spacex.com): Provides cargo and crew transportation services to the ISS.

Robotic Exploration of the Solar System

NASA and other space agencies greatly expanded robotic exploration during this period:

- Mars Exploration: Multiple orbiters, landers, and rovers, including Spirit, Opportunity, and Curiosity, have studied the Red Planet's geology and potential for past or present life.
- Outer Planets: The Voyager probes explored the gas giants and continue to provide data from interstellar space. Cassini-Huygens studied Saturn and its moons, while Juno currently orbits Jupiter.
- Inner Solar System: MESSENGER orbited Mercury, while Venus Express and Akatsuki studied Venus.
- Small Bodies: Missions like Dawn explored asteroids Vesta and Ceres, while Rosetta and its Philae lander studied comet 67P/Churyumov–Gerasimenko.
- Solar Observation: Spacecraft like SOHO and Parker Solar Probe have advanced our understanding of the Sun.

Significant companies involved in robotic exploration include:

1. Jet Propulsion Laboratory (www.jpl.nasa.gov): NASA's lead center for robotic exploration, responsible for missions like Mars rovers and Voyager.
2. Lockheed Martin (www.lockheedmartin.com/en-us/capabilities/space/exploration.html): Built spacecraft for various NASA missions, including OSIRIS-REx and InSight.
3. Airbus Defence and Space (www.airbus.com/en/products-services/space): Developed spacecraft for ESA missions like ExoMars and BepiColombo.
4. Maxar Technologies (www.maxar.com): Provides advanced space robotics and satellite systems for various missions.

These robotic missions have greatly expanded our knowledge of the solar system, providing insights into planetary formation, the potential for extraterrestrial life, and the dynamics of our cosmic neighborhood.

Here is some additional information with company names and URLs for the various areas of robotic space exploration:

Mars Exploration:
- NASA Jet Propulsion Laboratory (JPL) (https://www.jpl.nasa.gov/): Led development of Spirit, Opportunity, Curiosity, and Perseverance rovers
- Lockheed Martin (https://www.lockheedmartin.com/): Built Mars Reconnaissance Orbiter and InSight lander
- Airbus Defense and Space (https://www.airbus.com/space): Developed ExoMars Trace Gas Orbiter for ESA

Outer Planets:
- NASA JPL: Managed Voyager, Cassini, and Juno missions
- Johns Hopkins Applied Physics Laboratory (APL) (https://www.jhuapl.edu/): Built New Horizons spacecraft for Pluto flyby
- European Space Agency (ESA) (https://www.esa.int/): Developed Huygens probe for Titan landing

Inner Solar System:
- Johns Hopkins APL: Developed MESSENGER orbiter for Mercury
- ESA: Built Venus Express orbiter
- Japan Aerospace Exploration Agency (JAXA) (https://global.jaxa.jp/): Created Akatsuki Venus orbiter

Small Bodies:
- NASA JPL: Managed Dawn mission to Vesta and Ceres
- ESA: Led Rosetta mission to comet 67P
- DLR (German Aerospace Center) (https://www.dlr.de/EN/): Developed Philae lander for Rosetta

Solar Observation:
- NASA Goddard Space Flight Center (https://www.nasa.gov/goddard): Managed development of Parker Solar Probe
- ESA: Led SOHO mission in collaboration with NASA
- United Launch Alliance (https://www.ulalaunch.com/): Provided launch services for Parker Solar Probe

The post-Apollo era has been characterized by a shift from high-profile crewed missions to a more diverse approach, balancing human spaceflight with robotic exploration and international cooperation. This strategy has yielded significant scientific discoveries and technological advancements, paving the way for future exploration of the Moon, Mars, and beyond.

THE NEW AGE OF SPACE TECHNOLOGY

The dawn of the 21st century has ushered in a transformative era in space exploration, often referred to as "The New Age of Space Technology." This period is marked by a shift from government-dominated space activities to a dynamic ecosystem of private companies working alongside national space agencies. Technological advancements, private investment, and evolving government policies are driving this new age, making space more accessible and opening up unprecedented opportunities for exploration and commercialization.

Technological Innovations

Technological advancements are at the heart of the New Age of Space Technology. Key innovations include:

- Reusable Rocket Technology: Companies like SpaceX and Blue Origin have developed reusable rockets, significantly reducing the cost of access to space.
- 3D Printing: Relativity Space and other companies are using 3D printing to manufacture rocket components, reducing production time and costs.
- Advanced Materials and Propulsion Systems: Innovations in materials science and propulsion technology are enhancing the performance and efficiency of spacecraft.
- In-Orbit Servicing and Space Debris Removal: Companies like Astroscale (www.astroscale.com) are developing technologies to service satellites in orbit and mitigate space debris.

Commercial Applications

The commercialization of space is opening up new opportunities across various sectors:
- Satellite Constellations: Companies like SpaceX (Starlink) and OneWeb (www.oneweb.world) are deploying large constellations of satellites to provide global internet coverage.
- Earth Observation and Remote Sensing: Companies like Planet Labs (www.planet.com) and Maxar Technologies (www.maxar.com) are leveraging satellite technology for Earth observation and data analytics.
- Space Tourism: Virgin Galactic and Blue Origin are leading the way in making space tourism a reality.
- In-Space Manufacturing and Research: The ISS and private companies are exploring the potential of manufacturing and conducting research in microgravity environments.

Government and Private Sector Collaboration

Collaboration between government agencies and private companies is a hallmark of the New Age of Space Technology:

- NASA's Commercial Crew and Resupply Services: NASA partners with companies like SpaceX and Boeing (www.boeing.com/space) to transport crew and cargo to the ISS.
- International Partnerships: The Artemis program, led by NASA (www.nasa.gov/specials/Artemis), involves international partners and commercial companies in the effort to return humans to the Moon and establish a sustainable presence.

Future Prospects

The future of space exploration holds exciting possibilities:

- Mars Exploration and Colonization: SpaceX's Starship and NASA's Mars missions aim to explore and eventually colonize Mars.
- Deep Space Exploration and Interstellar Missions: Missions like NASA's Voyager and future interstellar probes aim to explore beyond our solar system.
- Space-Based Solar Power: The concept of harnessing solar power in space and transmitting it to Earth is being explored as a potential solution for sustainable energy.

Challenges and Considerations

While the New Age of Space Technology offers immense opportunities, it also presents significant challenges:

- **Space Debris and Orbital Congestion:** The increasing number of satellites and space activities raises concerns about space debris and the risk of collisions.
- **Legal and Regulatory Frameworks**: The current global legal framework for space activities needs to be updated to address emerging challenges and ensure responsible behavior in space.
- **Ethical Considerations:** The commercialization of space raises ethical questions about the exploitation of space resources and the potential impact on future generations.
- **Environmental Impacts**: The environmental impact of increased launch activities and space operations needs to be carefully managed.

The New Age of Space Technology represents a paradigm shift in how humanity approaches space exploration and utilization. With the combined efforts of innovative private companies and established space agencies, we are witnessing the birth of a new era that promises to expand our presence beyond Earth and unlock the vast potential of space for the benefit of all humankind.

"In the vastness of space, we discover not just new worlds,
but the limitless potential within ourselves."

THE IMPACT OF SPACE EXPLORATION

Space exploration has had a profound impact on various aspects of human life, from scientific discoveries and technological advancements to economic benefits and cultural inspiration. This section explores these impacts in detail, highlighting the contributions of significant players in the field.

Scientific Discoveries and Technological Advancements
Space exploration has led to numerous scientific breakthroughs and technological innovations that have significantly advanced our understanding of the universe and improved life on Earth.

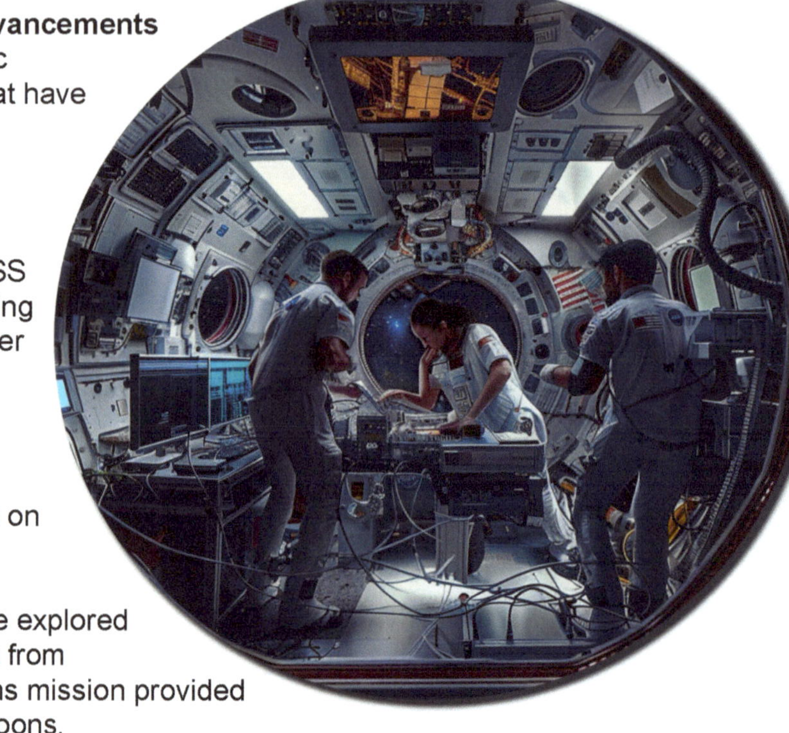

1. Scientific Discoveries:

- International Space Station (ISS): The ISS has been a hub for scientific research, leading to discoveries such as the fifth state of matter (Bose-Einstein condensate), methods to combat muscle atrophy and bone loss, and insights into human health in microgravity.
- Mars Exploration: Missions like NASA's Curiosity rover have provided valuable data on Mars' geology and climate, enhancing our understanding of the Red Planet.
- Outer Planets: The Voyager probes have explored the outer planets and continue to send data from interstellar space, while the Cassini-Huygens mission provided detailed information about Saturn and its moons.

2. Technological Advancements:

- **Reusable Rocket Technology:** Companies like SpaceX (www.spacex.com) and Blue Origin (www.blueorigin.com) have developed reusable rockets, significantly reducing the cost of access to space.
- **3D Printing:** Relativity Space (www.relativityspace.com) is pioneering the use of 3D printing to manufacture rockets, reducing production time and costs.
- **Advanced Materials:** Innovations in materials science, such as carbon fiber and advanced composites, have improved the performance and efficiency of spacecraft.

ECONOMIC BENEFITS AND SPIN-OFF TECHNOLOGIES

The economic impact of space exploration extends beyond the space industry, driving growth in various sectors through technological spin-offs and job creation.

1. Economic Growth:
- Job Creation: NASA's activities alone supported over 339,600 jobs nationwide and generated nearly $7.7 billion in federal, state, and local taxes in fiscal year 2021.
- Commercial Space Economy: The global space economy was valued at $469 billion in 2021, with significant contributions from satellite services, space tourism, and in-space manufacturing.

2. Spin-Off Technologies:
- Medical Innovations: Space research has led to advancements in medical technologies, such as improved water purification systems, drug development using protein crystals, and methods to combat muscle atrophy and bone loss.
- Consumer Products: Technologies developed for space missions have been adapted for everyday use, including CMOS image sensors in digital cameras, memory foam, and scratch-resistant sunglasses.
- Environmental Monitoring: Satellites provide critical data for monitoring climate change, pollution, and natural disasters, aiding in environmental protection and disaster response.

INSPIRATION AND CULTURAL IMPACT

Space exploration has inspired generations of scientists, engineers, and the general public, fostering a culture of innovation and curiosity.

Educational Inspiration:
- STEM Education: Space missions, such as the Apollo moon landings and the Mars rovers, have inspired countless students to pursue careers in science, technology, engineering, and mathematics (STEM).
- Public Engagement: Programs like NASA's TechRise Student Challenge and educational centers like the Earth and Space Expedition Center in Phoenix engage the public and promote interest in space exploration.

Cultural Impact:

1. Global Perspective: The iconic images of Earth from space, such as the "Blue Marble" photograph, have fostered a sense of global unity and environmental awareness.

2. Art and Media: Space exploration has profoundly influenced art, literature, and media, inspiring works that capture the imagination and convey the wonder of the cosmos.

"Space exploration is the compass that guides our species towards its greatest adventure."

SCIENCE FICTION'S INFLUENCE ON CULTURE

Science fiction has played a pivotal role in shaping public perception and enthusiasm for space exploration. Two of the most influential franchises, Star Trek and Star Wars, have not only entertained millions but also inspired technological advancements and cultural shifts.

STAR TREK: A VISIONARY INFLUENCE

Star Trek, created by Gene Roddenberry, debuted in 1966 and quickly became a cultural touchstone. Its optimistic vision of the future, where humanity has transcended its differences to explore the cosmos in peace and cooperation, has inspired generations of scientists, engineers, and space enthusiasts.

• Technological Inspiration: Many of the gadgets and concepts from Star Trek have influenced real-world technology. For instance, the flip phone was inspired by the communicator, and voice-activated computers echo the ship's computer. The franchise's portrayal of advanced medical devices, such as the tricorder, has also spurred innovation in medical technology.

• Diversity and Inclusion: Star Trek broke new ground with its diverse cast, featuring characters from different racial and ethnic backgrounds working together as equals. This was a radical statement during the 1960s and has continued to influence social attitudes and inspire inclusivity in various fields, including space exploration.

• NASA and Star Trek: The relationship between NASA and Star Trek is well-documented. NASA named the first Space Shuttle "Enterprise" after a fan-led campaign, and several astronauts, including Mae Jemison, have cited the series as an inspiration for their careers. Jemison even appeared in an episode of Star Trek: The Next Generation.

STAR WARS: A CULTURAL PHENOMENON

Star Wars, created by George Lucas, premiered in 1977 and revolutionized the entertainment industry. Its epic storytelling, groundbreaking special effects, and memorable characters have left an indelible mark on popular culture.

• Technological Impact: Star Wars has influenced the development of real-life space technology. For example, NASA has tested laser communication systems inspired by the franchise's depiction of advanced technology. The Strategic Defense Initiative, proposed in the 1980s, was popularly nicknamed "Star Wars" due to its futuristic defense concepts.

• Cultural Lexicon: Phrases like "May the Force be with you" and characters such as Darth Vader have become part of the global cultural lexicon. The franchise's themes of good versus evil, heroism, and adventure resonate with audiences worldwide, fostering a shared cultural experience.

SPACEPOSIUM

- Inspiring Space Exploration: The imaginative worlds and technologies depicted in Star Wars have inspired many to pursue careers in science and engineering. SpaceX founder Elon Musk, for instance, named his Falcon rockets after the Millennium Falcon, highlighting the franchise's influence on modern space endeavors.

BROADER IMPACT OF SCIENCE FICTION

Beyond Star Trek and Star Wars, other science fiction works have also played significant roles in inspiring space exploration and shaping cultural attitudes.

- 2001: A Space Odyssey: Stanley Kubrick's 1968 film presented a realistic and awe-inspiring vision of space travel, influencing both public perception and the aesthetics of space exploration.
- The Space Race and Pop Culture: The Space Race of the 1960s and 70s spurred a surge in space-themed media, from movies and TV shows to theme park attractions. This cultural shift helped maintain public interest in space exploration and inspired future generations to dream of the stars.

The cultural impact of science fiction, particularly through iconic franchises like Star Trek and Star Wars, has been profound. These stories have not only entertained but also inspired technological innovation, fostered inclusivity, and encouraged the pursuit of space exploration. As we continue to explore the cosmos, the legacy of these cultural phenomena will undoubtedly continue to inspire and shape our journey. By weaving together the dreams of the past with the technological advancements of the present, science fiction has helped humanity envision and strive for a future among the stars.

CULTURAL IMPACT OF THE "BLUE MARBLE" PHOTOGRAPH:
1. Environmental Awareness: The image played a crucial role in catalyzing the modern environmental movement by showcasing Earth's beauty and fragility. It helped launch the first Earth Day in 1970 and inspired greater environmental consciousness globally.
2. Global Unity: The photo fostered a sense of global unity by showing Earth without national borders, emphasizing our shared planet and humanity.
3. Philosophical Shift: It marked a turning point from faith in unlimited progress to understanding the limitations of our planetary environment. The image made Earth appear more unique and precious.
4. Artistic Inspiration: The "Blue Marble" has been widely reproduced in art, appearing on posters, paintings, stamps, and various media. It became an iconic image representing Earth.
5. Scientific Impact: The photo contributed to the development of "Earth systems science" and influenced how we study our planet as an interconnected system.
6. Popular Culture: The image has been featured on magazine covers, books, and various products, becoming a recognizable symbol in popular culture.

CULTURAL IMPACT OF HUBBLE SPACE TELESCOPE IMAGES:

1. Public Engagement: Hubble's stunning images have captivated the public imagination, making astronomy and space science more accessible and exciting to the general public.
2. Art and Design: Hubble images have inspired various forms of art and design, appearing on clothing, accessories, and even in tattoos.
3. Music: Composers have created symphonies and other musical works inspired by Hubble images and the sense of wonder they evoke.
4. Literature: Numerous books have been written about Hubble's discoveries and its impact on our understanding of the universe.
5. Education: Hubble images are widely used in educational materials, helping to teach astronomy and inspire interest in science among students.
6. Cultural Icons: Hubble and its images have become cultural icons, appearing on stamps, coins, and in popular media.
7. Architectural and Interior Design: Hubble images have been incorporated into architectural designs and used as art in public spaces and museums.
8. Digital Media: Hubble images are frequently used in digital art, screensavers, and online content, spreading their impact through modern technology.

Both the "Blue Marble" photograph and Hubble images have significantly influenced how we perceive our place in the universe, inspiring awe, scientific curiosity, and a sense of cosmic perspective across various aspects of culture.

RECENT IMPACT OF THE JAMES WEBB SPACE TELESCOPE:

The James Webb Space Telescope (JWST), launched in December 2021, has reignited public interest in space exploration and astronomy. Its stunning images of distant galaxies and exoplanets have captured the imagination of people worldwide, much like science fiction has done for decades. The JWST's ability to peer further into space and time than ever before is turning some of science fiction's wildest dreams into reality.

Key impacts include:
1. Inspiring a new generation: The JWST's discoveries are inspiring young people to pursue careers in STEM fields, much like the Apollo missions did in the 1960s.
2. Bridging science and popular culture: Images from JWST have gone viral on social media, bringing complex astronomical concepts to the general public in an accessible way.
3. Fueling new sci-fi narratives: The telescope's findings are already influencing science fiction writers, providing new settings and concepts for stories about deep space and alien worlds.
4. Validating sci-fi concepts: Some of JWST's observations, such as detecting water on exoplanets, are bringing long-standing science fiction ideas closer to reality.

BROADER GLOBAL SCI-FI THEMES AND MOVIES:

Science fiction is a global phenomenon, with different cultures contributing unique perspectives:
1. Afrofuturism: This cultural aesthetic combines science fiction with African and African diaspora cultures. Films like "Black Panther" (2018) showcase advanced African civilizations, challenging Western-centric sci-fi narratives.
2. Chinese Sci-Fi: Authors like Liu Cixin (The Three-Body Problem) and films such as "The Wandering Earth" (2019) blend traditional Chinese philosophy with futuristic concepts, often exploring themes of collective action and long-term planning.
3. Indian Sci-Fi: Bollywood has produced sci-fi films like "PK" (2014) and "2.0" (2018) that often incorporate social commentary on Indian society alongside futuristic elements.
4. Russian Sci-Fi: With a rich literary tradition including authors like Stanislaw Lem, Russian sci-fi often explores philosophical themes. Films like "Sputnik" (2020) continue this tradition in the modern era.
5. Japanese Sci-Fi: Anime and manga have long been at the forefront of science fiction, with works like "Akira" and "Ghost in the Shell" influencing global pop culture and exploring themes of technology's impact on humanity.
6. European Sci-Fi: Films like "High Life" (2018) and "Aniara" (2018) offer contemplative, often bleaker visions of the future, focusing on existential questions and the human condition in space.

These diverse perspectives in science fiction reflect global concerns about technology, the environment, and the future of humanity while also showcasing cultural values and philosophies. The genre continues to evolve, influenced by real-world scientific advancements like the James Webb Space Telescope, creating a feedback loop between imagination and reality that pushes both science and storytelling forward.

ASPIRATIONS FOR THE FUTURE.

Key Players in the New Space Age

The rise of commercial space companies has been a defining feature of this new era. These enterprises are driving innovation, reducing costs, and expanding access to space. Key players include:

1. SpaceX (www.spacex.com)

- Reusable Rocket Technology: SpaceX has revolutionized the launch industry with its reusable Falcon 9 and Falcon Heavy rockets. The development of the fully reusable Starship aims to enable deep space exploration and Mars colonization.

- Commercial Crew and Cargo Missions: SpaceX provides crew and cargo transportation services to the International Space Station (ISS) under NASA's Commercial Crew and Commercial Resupply Services programs.

2. Blue Origin (www.blueorigin.com)

- New Shepard: A reusable suborbital vehicle designed for space tourism and research.

- New Glenn: An orbital launch vehicle aimed at providing reliable access to space for a variety of missions.

Long-term Vision: Blue Origin is committed to building a future in which millions of people live and work in space.

3. Virgin Galactic (www.virgingalactic.com) and Virgin Orbit (www.virginorbit.com)

- Space Tourism: Virgin Galactic focuses on suborbital space tourism with its SpaceShipTwo spaceplane.

- Small Satellite Launches: Virgin Orbit provides launch services for small satellites using its air-launched Launcher One rocket.

4. Rocket Lab (www.rocketlabusa.com)

- Electron Rocket: This company specializes in small satellite launches and is working on making the Electron rocket partially reusable.

- Future Missions: Plans include missions to Venus and lunar exploration.

5. Relativity Space (www.relativityspace.com)

- 3D-Printed Rockets: Pioneering the use of 3D printing to manufacture rockets, aiming to revolutionize spacecraft production.

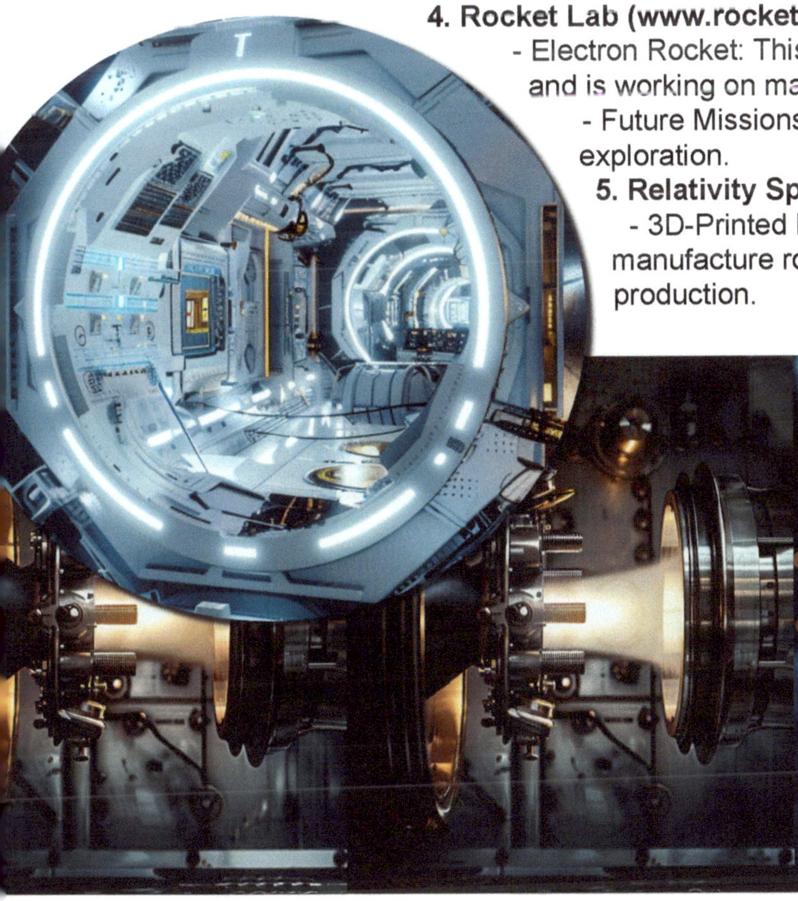

SPACEPOSIUM

Conclusion

The New Age of Space Technology represents a remarkable evolution of humanity's enduring fascination with the cosmos. From the ancient astronomers who meticulously charted the heavens to today's cutting-edge space exploration, our journey of discovery has been a testament to human ingenuity and perseverance.

This new era is not merely a continuation of past efforts but a transformative age that builds upon millennia of astronomical knowledge and technological progress. The collaborative efforts of government agencies and private companies are propelling us forward, creating significant scientific, economic, and cultural impacts that echo the profound influence space has always had on human civilization.

As we venture further into space, we carry with us the legacy of countless generations who looked up at the night sky with wonder. The ancient Egyptians who aligned their pyramids with the stars, the Mayans who crafted intricate calendars based on celestial movements, and the Greek philosophers who pondered the nature of the cosmos all contribute to the foundation upon which our modern space endeavors stand.

The advancements in space technology are not just about reaching new frontiers; they enhance life on Earth through technological spin-offs, economic growth, and by inspiring the next generation of innovators – much as the mysteries of the night sky inspired our ancestors. As we unlock the vast potential of space for the benefit of all humankind, we continue a journey of exploration and discovery that has been central to human experience for thousands of years.

In this epic quest, we are not just exploring space; we are redefining what it means to be human, pushing the boundaries of our capabilities, and forging a future where the stars are within our reach. The New Age of Space Technology is both a beacon of hope and a symbol of our collective aspirations, heralding a new chapter in the grand narrative of human achievement – a narrative that stretches back to the very dawn of civilization and now extends beyond our planet.

The cultural impact of science fiction, particularly through iconic franchises like Star Trek and Star Wars, has been profound. These stories have not only entertained but also inspired technological innovation, fostered inclusivity, and encouraged the pursuit of space exploration. As we continue to explore the cosmos, the legacy of these cultural phenomena will undoubtedly continue to inspire and shape our journey.

As we stand on the cusp of interplanetary exploration and the potential colonization of other worlds, we do so as inheritors of an ancient legacy of cosmic curiosity. The New Age of Space Technology is thus not just about the future; it is the culmination of humanity's age-old dream to understand and explore the universe.

ENVISIONING TOMORROW'S POTENTIAL

Envisioning Tomorrow's Potential in spacecraft design reveals a fascinating blend of realism and futurism, especially when considering the potential acceleration of technological progress through AI.

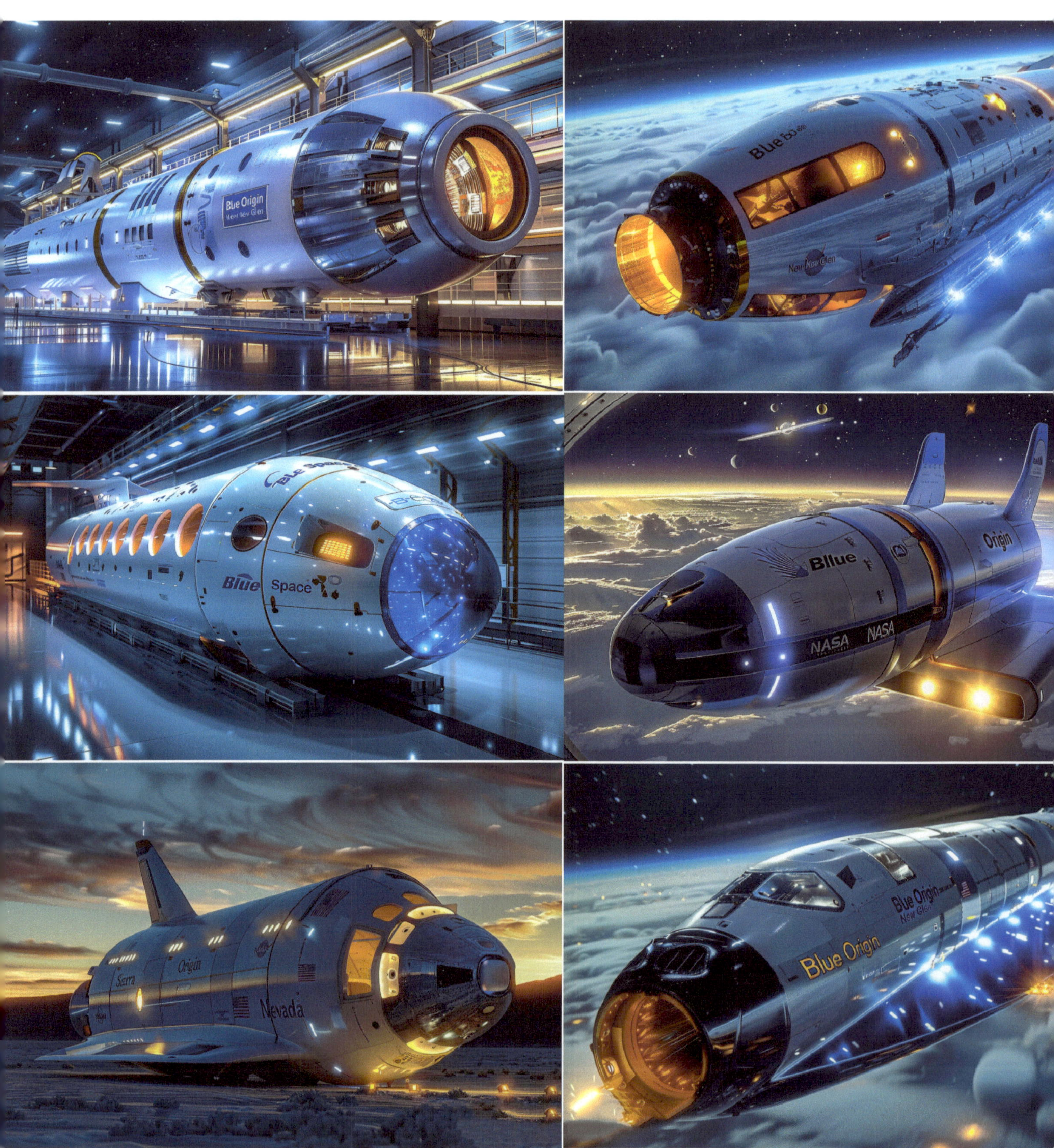

Let's take a quick "AI Leap" and explore some spacecraft concepts that could emerge in the near or distant future:

Modular Spherical Habitats
Rather than the traditional cylindrical modules seen in current space stations, future spacecraft may feature interconnected spherical habitats. These blobby, potato-like designs could be more efficient for maintaining internal pressure and maximizing usable space. The spherical sections, welded together in various sizes, would create a unique, organic appearance while serving practical purposes for long-duration space missions.

Sensor-Driven Navigation
Future spacecraft might lack traditional windows or cockpits, instead relying on advanced sensor arrays and AI-driven navigation systems. Crew members could use virtual reality interfaces or direct neural links to "see" in all directions around the ship, eliminating the need for a bridge at the front. This design would allow for more flexible internal layouts and improved protection from space debris.

Asymmetrical Propulsion Systems
Moving away from the symmetrical designs common in science fiction, realistic future spacecraft might have asymmetrical shapes dictated by their propulsion systems and fuel storage. Different propellant densities and innovative engine designs could lead to unconventional, yet highly efficient spacecraft configurations that optimize mass distribution and thrust capabilities.

Adaptive Outer Hulls
Envision spacecraft with smart, re-configurable outer surfaces that can adapt to various space environments. These hulls might incorporate self-healing materials, adjustable heat shields, and metamorphic structures that can alter the ship's aerodynamics for potential atmospheric entry on distant planets.

Artificial Gravity Systems
While not relying on the classic "spinning wheel" design, future spacecraft could incorporate more subtle artificial gravity systems. These might involve localized gravitational fields generated by advanced technology, allowing for Earth-like conditions in specific areas of the ship without compromising the overall design flexibility.

By combining these elements, we can imagine spacecraft that are both pragmatic and visually striking, pushing the boundaries of what's possible while remaining grounded in plausible technological advancements.

CHAPTER II
SPACE TECHNOLOGY

Space technology can be defined as the application of scientific knowledge, engineering principles, and technological innovations to develop systems, devices, and capabilities for use in space exploration, space-based operations, and Earth applications derived from space activities.

Space technology is a multidisciplinary field that integrates various scientific and engineering disciplines, including aerospace engineering, astrophysics, computer science, materials science, and many others.

This collaborative approach leverages the expertise from different areas to develop systems, devices, and capabilities for space exploration and operations. By combining knowledge from these diverse fields, space technology enables the design, development, and operation of launch vehicles, spacecraft, satellites, and other space-based systems. This interdisciplinary effort not only advances our understanding of the universe but also leads to innovations that benefit life on Earth, such as advancements in medicine, environmental monitoring, and telecommunications.

There are several compelling reasons why people should try to understand space technology:

1. Inspiration and innovation: Space exploration captures our imagination and inspires us to push the boundaries of human knowledge and capability. Understanding space technology can spark creativity and drive innovation across various fields.

2. Educational benefits: Space technology serves as a powerful motivator for people to engage with STEM (Science, Technology, Engineering, and Mathematics) fields. It encourages critical thinking and problem-solving skills that are valuable in many areas of life and work.

3. Technological advancements: Many technologies developed for space exploration have practical applications on Earth. Understanding space technology can help us appreciate and potentially contribute to innovations in areas like communications, navigation (GPS), weather forecasting, and medical advancements.

4. Global challenges: Space technology plays a crucial role in addressing global issues such as climate change, disaster management, and environmental monitoring. Understanding these technologies can help us better comprehend and contribute to solving these challenges.

5. Economic opportunities: The space sector is a growing industry that creates jobs and drives economic growth. Understanding space technology can open up career opportunities and help individuals participate in this expanding field.

6. Perspective on our place in the universe: Learning about space technology and exploration provides a broader perspective on Earth's place in the cosmos, fostering a sense of global unity and environmental stewardship.

7. International cooperation: Space exploration often involves collaboration between nations, promoting peaceful international relations and shared scientific goals.

8. Future preparedness: As we look towards potential colonization of other planets or mining of space resources, understanding space technology will be crucial for future generations.

9. Enhancing daily life: Many space technologies have direct applications in our everyday lives, from satellite communications to Earth observation data used in agriculture and urban planning.

10. Scientific discovery: Space technology enables us to make groundbreaking discoveries about the universe, our solar system, and even our own planet, expanding our knowledge and understanding of the cosmos.

By understanding space technology, individuals can become more informed citizens, potentially contribute to scientific and technological advancements, and gain a deeper appreciation for the role of space exploration in shaping our present and future.

Specifically:

1. It encompasses the design, development, manufacturing, and operation of:
- Launch vehicles and spacecraft
- Satellites and space stations
- Scientific instruments and sensors for space-based research
- Communication and navigation systems
- Life support systems for human spaceflight
- Robotic systems for space exploration

2. It includes technologies that enable:
 - Space travel and exploration beyond Earth's atmosphere
    - Earth observation and remote sensing
    - Satellite communications and navigation
    - Space-based scientific research and experimentation

3. It also covers the development of:
 - Advanced materials and manufacturing techniques for space applications
    - Propulsion systems for spacecraft
    - Power generation and storage systems for space use
    - Thermal management and radiation protection technologies
    - Software and AI systems for space operations

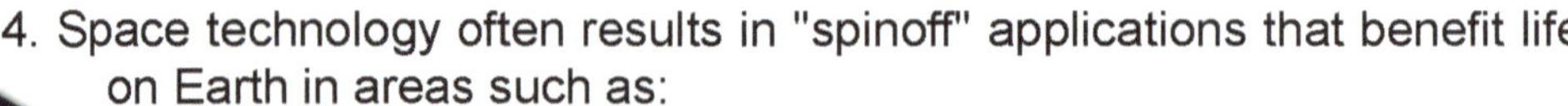

4. Space technology often results in "spinoff" applications that benefit life on Earth in areas such as:
- Medicine and healthcare
- Environmental monitoring and protection
- Transportation and navigation
- Computing and telecommunications

In essence, space technology refers to any technological advancement or tool created specifically for use in outer space or derived from space-related research and development efforts. It plays a crucial role in advancing our understanding of the universe, improving life on Earth, and expanding human presence beyond our home planet.

ROCKET SCIENCE: THE VEHICLES OF SPACE TRAVEL

Rocket science is the foundation of space exploration, enabling us to overcome Earth's gravity and venture into the cosmos. This section explores the fundamental principles of rocketry and the various types of launch vehicles used in space exploration.

SUSTAINABLE JOURNEYS: REUSABLE LAUNCH SYSTEMS

Reusable launch systems are changing the economics of space travel by allowing rockets to be used multiple times, significantly reducing costs.

Notable companies in this area include:

- SpaceX (https://www.spacex.com): Pioneered reusable rocket technology with their Falcon 9 and Falcon Heavy rockets.
- Blue Origin (https://www.blueorigin.com): Developing the New Glenn reusable orbital launch vehicle.
- Rocket Lab (https://www.rocketlabusa.com): Working on making their Electron rocket partially reusable.

Reusable rockets work by landing their first stage boosters back on Earth after launch, either on land or on floating platforms at sea. This allows the most expensive part of the rocket to be refurbished and reused, much like how airplanes are reused for multiple flights.

THE ART OF SPACE NAVIGATION

Space navigation involves guiding spacecraft through the vast emptiness of space to reach their intended destinations accurately.

Key players in space navigation include:

- NASA (https://www.nasa.gov): Develops advanced navigation technologies for deep space missions.
- European Space Agency (https://www.esa.int): Works on navigation systems for various space missions.
- Lockheed Martin (https://www.lockheedmartin.com): Provides navigation systems for many spacecraft and satellites.

Space navigation relies on a combination of technologies, including star trackers (which use the positions of stars for orientation), inertial measurement units (which detect changes in a spacecraft's motion), and communication with ground-based tracking stations. These systems work together to ensure spacecraft stay on course during their long journeys through space.

SMART EXPLORERS: THE ROLE OF AI AND ROBOTICS IN SPACE

Artificial Intelligence (AI) and robotics are playing an increasingly important role in space exploration, allowing for more autonomous and efficient missions. Companies and organizations at the forefront of this field include:

- NASA JPL (https://www.jpl.nasa.gov): Develops AI systems for Mars rovers and other robotic explorers.
- Maxar Technologies (https://www.maxar.com): Creates advanced robotic systems for satellite servicing and space exploration.
- Astrobotic (https://www.astrobotic.com): Develops autonomous lunar landers and rovers.

AI and robotics in space can perform tasks ranging from autonomous navigation on planetary surfaces to complex decision-making in unpredictable environments. For example, Mars rovers use AI to identify interesting geological features and plan their own routes, while robotic arms on satellites can perform repairs and maintenance in orbit.

COSMIC CONNECTIONS: SPACE COMMUNICATIONS

Space communications systems are crucial for maintaining contact with spacecraft and transmitting data back to Earth. Space communications involve transmitting radio signals over vast distances, often using large dish antennas on Earth to communicate with spacecraft equipped with smaller antennas. These systems allow us to control spacecraft, receive scientific data and images, and even communicate with astronauts in space.

POWERING THE COSMOS: SPACE POWER SYSTEMS

Reliable power systems are essential for spacecraft to operate in the harsh environment of space. Notable organizations in this field include:

- NASA Glenn Research Center (https://www.nasa.gov/glenn): Develops advanced power systems for space applications.
- Boeing (https://www.boeing.com): Designs and manufactures solar arrays and other power systems for satellites and spacecraft.
- Airbus Defence and Space (https://www.airbus.com/space): Produces solar arrays and power systems for various space missions.

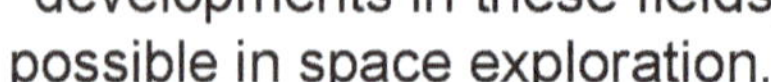

Space power systems typically rely on solar panels to generate electricity from sunlight. For missions traveling far from the Sun, where solar power is less effective, radioisotope thermoelectric generators (RTGs) are used. These devices convert heat from the decay of radioactive material into electricity, providing consistent power for decades.

Each of these areas plays a crucial role in enabling humanity's exploration of space, from the powerful rockets that lift us off the Earth to the intricate systems that keep our spacecraft functioning in the harsh environment of space. As technology continues to advance, we can expect even more exciting developments in these fields, pushing the boundaries of what's possible in space exploration.

ROCKET SCIENCE: THE VEHICLES OF SPACE TRAVEL

As we look to the future of space exploration and launch capabilities, two key emerging technologies stand out: methane-fueled engines and aerospike engines. These innovations promise to revolutionize rocket propulsion, offering improved efficiency, reusability, and performance across a wide range of altitudes.

METHANE-FUELED ENGINES

Methane (CH_4) is gaining traction as a rocket fuel due to several advantages:

1. Reusability: Methane burns cleaner than traditional kerosene (RP-1), resulting in less coking and soot buildup. This makes methane-fueled engines easier to refurbish and reuse multiple times.
2. Mars missions: Methane can potentially be produced on Mars using in-situ resource utilization (ISRU) techniques, making it an attractive fuel for future Mars missions.
3. Efficiency: Methane offers a higher specific impulse than kerosene, potentially providing a 20% performance benefit when factoring in increased efficiency.
4. Storage: Methane's storage temperature is closer to that of liquid oxygen, simplifying tank design and reducing insulation requirements.
5. Cost: Industrial-grade liquefied natural gas (LNG) can be used as fuel, which is more readily available and less expensive than highly refined RP-1 kerosene.

AEROSPIKE ENGINES

Aerospike engines offer a promising alternative to traditional bell nozzles, with several key advantages:

1. Altitude compensation: Aerospike engines maintain their efficiency across a wide range of altitudes, unlike traditional bell nozzles which are optimized for a specific altitude.
2. Compact design: The aerospike nozzle can be packaged more compactly than a traditional bell nozzle, potentially reducing overall rocket size.
3. Thrust vectoring: Aerospike engines can use pressure injection for thrust vectoring, eliminating the need for heavy gimbal systems used in traditional rockets.
4. Efficiency: Aerospike engines promise increased efficiency throughout ascent due to their inherent altitude compensation.
5. Single-stage-to-orbit potential: The efficiency gains of aerospike engines make them attractive for potential single-stage-to-orbit (SSTO) vehicle designs.

DUAL EXPANDER AEROSPIKE NOZZLE (DEAN)

An innovative concept combining the benefits of expander cycle engines and aerospike nozzles is the Dual Expander Aerospike Nozzle (DEAN):

- The DEAN design uses both fuel and oxidizer to cool different parts of the engine, maximizing the use of propellants for cooling.
- This approach potentially allows for higher thrust expander cycle engines, overcoming traditional limitations.
- The Air Force Institute of Technology (AFIT) has conducted research on methane-fueled DEAN concepts.

FUTURE PROSPECTS

As these technologies mature, we can expect to see:

1. Increased focus on reusability and cost reduction in launch vehicles.
2. More efficient upper stages for deep space exploration.
3. Potential breakthroughs in single-stage-to-orbit vehicle designs.
4. Improved capabilities for Mars missions and in-situ resource utilization.

The AFRL's ARISE (Aerospike Rocket Integration and Suborbital Experiment) project aims to demonstrate a modular aerospike rocket engine in flight, potentially marking a significant milestone in aerospike engine development.

Methane-fueled engines and aerospike nozzles represent two of the most promising advancements in rocket propulsion technology. As manufacturing techniques improve and more real-world testing is conducted, these technologies have the potential to significantly impact the future of space exploration and launch vehicle design. The coming years will likely see increased investment and development in these areas, potentially leading to more efficient, capable, and cost-effective access to space.

BASIC PRINCIPLES OF ROCKET PROPULSION

Rocket propulsion is based on Newton's Third Law of Motion: for every action, there is an equal and opposite reaction. In rockets, this principle is applied through the expulsion of high-speed gasses in one direction, which propels the rocket in the opposite direction.

Key concepts in rocket propulsion include:

- **Thrust:** Thrust is the force that propels a rocket through space. It's generated by the rocket engine as it expels mass (propellant) in one direction, causing the rocket to move in the opposite direction, in accordance with Newton's Third Law of Motion.
- **Specific Impulse (Isp):** Specific impulse is a measure of how efficiently a rocket uses propellant to produce thrust. It's defined as the total impulse (thrust integrated over time) delivered per unit of propellant consumed. In simpler terms, it's the amount of thrust produced per unit of propellant flow rate.
- **Mass Ratio:** The mass ratio is a critical parameter in rocket design, directly related to the rocket's performance.

These three concepts - thrust, specific impulse, and mass ratio - are fundamental to understanding and designing rocket propulsion systems. They interact in complex ways: for instance, a higher thrust often comes at the cost of lower specific impulse, and achieving a high mass ratio requires careful engineering to minimize the mass of the rocket structure and maximize the propellant fraction.

Types of Rocket Engines

1. Liquid-fuel rockets: Use liquid propellants, typically a fuel and an oxidizer.
2. Solid-fuel rockets: Use a solid propellant mixture.
3. Hybrid rockets: Combine aspects of both liquid and solid-fuel systems.

Let's elaborate on the different types of rocket engines:

Liquid-fuel rockets:

- Liquid-fuel rockets use liquid propellants, typically a fuel and an oxidizer, stored separately and mixed in the combustion chamber.

Key features:
- Higher specific impulse (efficiency) than solid rockets
- Can be throttled, stopped, and restarted
- More complex due to pumps, valves, and plumbing
- Propellants can be stored at cryogenic temperatures for higher performance

Examples:
- SpaceX Merlin engine (kerosene/liquid oxygen)
- RS-25 Space Shuttle Main Engine (liquid hydrogen/liquid oxygen)

Solid-fuel rockets:

Solid-fuel rockets use a premixed solid propellant, typically in the form of a rubber-like material.

Key features:
- Simpler design with fewer moving parts
- Cannot be throttled or shut off once ignited
- Generally lower specific impulse than liquid rockets
- Can be stored for long periods, making them ideal for military applications

Examples:
- Space Shuttle Solid Rocket Boosters
- Minuteman intercontinental ballistic missile

Hybrid rockets:

Hybrid rockets combine aspects of both liquid and solid-fuel systems, typically using a solid fuel and a liquid or gaseous oxidizer.

Key features:
- Safer than pure solid or liquid rockets due to propellant separation
- Generally simpler than liquid rockets but more complex than solid rockets
- Often used in experimental and amateur rocketry

Examples:
- SpaceShipOne and SpaceShipTwo (rubber/nitrous oxide)
- Firefly Alpha (paraffin/liquid oxygen)
- Copenhagen Suborbitals HEAT rocket (polyurethane/liquid oxygen)

MAJOR COMPONENTS OF A LAUNCH VEHICLE

1. Stages: Most rockets use multiple stages to increase efficiency.
2. Payload fairing: Protects the payload during launch.
3. Propellant tanks: Store fuel and oxidizer.
4. Engines: Provide thrust for the rocket.
5. Guidance and control systems: Steer the rocket.

Now let's elaborate on the major components of a **Launch Vehicle:**

Stages:

Rockets use multiple stages to increase efficiency, and they are based on the principle of mass reduction during flight. As each stage burns its fuel and completes its function, it's jettisoned, reducing the overall mass of the vehicle. This allows subsequent stages to accelerate more efficiently. Typically, launch vehicles have two or three stages, but some may have more.

- First stage: Usually the largest and most powerful, designed to lift the rocket off the launch pad and accelerate it through the densest part of the atmosphere.
- Second stage: Takes over once the first stage is depleted and separated, continuing to accelerate the payload to orbital velocity.
- Upper stages: Some rockets have additional upper stages for reaching higher orbits or interplanetary trajectories.

Payload fairing:

The payload fairing, also known as the nose cone, is a protective shell that encapsulates the payload (satellite, spacecraft, etc.) during launch. Its primary functions are:

- Aerodynamic streamlining to reduce drag during ascent through the atmosphere
- Protection from aerodynamic forces, heating, and acoustic vibrations

Propellant tanks:

These store the fuel and oxidizer needed for the rocket engines. Key features include:

- Usually the largest structures in the rocket
- Often serve as part of the primary load-bearing structure
- Must withstand high pressures and extreme temperatures (especially for cryogenic propellants)
- Typically made of lightweight but strong materials like aluminum alloys or composite materials

Engines:

Rocket engines provide the thrust necessary to propel the vehicle. They come in various types:

- Liquid-propellant engines: Mix fuel and oxidizer in a combustion chamber
- Solid-propellant engines: Use a pre-mixed solid propellant
- Hybrid engines: Combine aspects of both liquid and solid engines
- Some rockets use multiple engines per stage for increased thrust and redundancy

Guidance and control systems:

These systems are responsible for steering the rocket and ensuring it follows the intended trajectory. They include:

- Inertial measurement units (IMUs) to track the rocket's position and orientation
- Flight computers to process data and make course corrections
- Control mechanisms like gimbaled engines, thrust vectoring, or small thrusters for attitude control
- Communication systems to receive commands from ground control and transmit telemetry data

Additional important components include:

- **Interstage structures:** Connect different stages and house separation mechanisms.
- **Avionics:** Electronic systems for communication, data handling, and mission management.
- **Thermal protection systems:** Protect the vehicle from extreme temperatures during ascent and, for some vehicles, re-entry.

- **Ground support equipment interfaces:** Connections for fueling, electrical power, and communication while on the launch pad.
- **Recovery systems (for reusable vehicles):** Parachutes, landing legs, or other mechanisms to enable safe return and reuse of certain components.

Each of these components plays a crucial role in the successful operation of a launch vehicle, working together as an integrated system to deliver payloads to their intended orbits or trajectories.

CURRENT AND HISTORICAL LAUNCH VEHICLES

Here are just a few of the current and historical launch vehicles mentioned:

Saturn V:

- Developed by NASA for the Apollo program in the 1960s and 1970s
- Remains the most powerful rocket ever successfully flown
- Three-stage liquid-fueled rocket, standing 363 feet (110.6 meters) tall
- Capable of launching 310,000 pounds (140,000 kg) to low Earth orbit
- Used for all crewed Apollo lunar missions and to launch Skylab
- Notable for its reliability, with 13 launches and no failures

The **Saturn V rocket** had a significant influence on future launch vehicle designs in several ways:

1. Multi-stage design: The Saturn V's three-stage design became a model for many subsequent heavy-lift rockets. This approach of using multiple stages that are jettisoned as fuel is depleted remains common in modern rockets.
2. Propulsion technology: The F-1 engines used on the Saturn V's first stage were the most powerful single-chamber liquid-fueled rocket engines ever developed. Their design influenced future large rocket engine development.
3. Use of liquid hydrogen fuel: The Saturn V's upper stages used liquid hydrogen fuel, which provided high efficiency. This helped establish liquid hydrogen as a preferred fuel for upper stages on many subsequent rockets.
4. Size and payload capacity: The Saturn V demonstrated the feasibility of very large rockets capable of lifting massive payloads. This paved the way for other heavy-lift vehicles like the Space Shuttle and modern rockets like SpaceX's Falcon Heavy and NASA's Space Launch System (SLS).
5. Reusability concepts: While the Saturn V itself was not reusable, its development led to studies on partially and fully reusable variants that influenced later reusable rocket designs.

Space Shuttle:

- NASA's partially reusable launch system, operational from 1981 to 2011
- Consisted of an orbiter vehicle, external tank, and two solid rocket boosters
- Orbiter and boosters were reusable, while the external tank was expended
- Could carry up to 27,500 kg (60,600 lb) to low Earth orbit
- Flew 135 missions, including deploying satellites and constructing the International Space Station
- Suffered two fatal accidents: Challenger (1986) and Columbia (2003)

Falcon 9:

- SpaceX's two-stage rocket, first launched in 2010
- First stage is reusable, capable of landing vertically on ground or drone ships
- Can carry up to 22,800 kg (50,265 lb) to low Earth orbit in expendable configuration
- Highly cost-effective due to its reusability
- Has flown over 200 successful missions as of 2023
- Used for satellite launches, cargo resupply to ISS, and crewed missions

Ariane 5:

- European Space Agency's heavy-lift launch vehicle, operational from 1996 to 2023
- Two main variants: Ariane 5 ECA for commercial payloads and Ariane 5 ES for government missions
- Could launch up to 21,000 kg (46,300 lb) to low Earth orbit
- Known for its reliability, with 117 launches and only 5 partial or total failures
- Used primarily for commercial satellite launches and some high-profile scientific missions like the James Webb Space Telescope
- Recently retired, to be replaced by the Ariane 6
- Each of these launch vehicles represents a significant era in space exploration and satellite deployment, showcasing the evolution of rocket technology from the early days of the Space Race to the current era of commercial spaceflight and reusability.

As we conclude our exploration of space technology, we find ourselves on the brink of a new era in human exploration. The principles and innovations we've examined are not mere facts and figures; they are the foundation of our cosmic future. From the powerful launch of the Saturn V to the elegant landing of the Falcon 9, each vehicle symbolizes human ingenuity and our highest aspirations.

The journey through rocket science is a testament to human achievement. Inspired by visionaries like Tsiolkovsky and Goddard and realized by countless dedicated engineers and scientists, each component, from the fiery core of rocket engines to the precise choreography of guidance systems, contributes to a grand symphony of exploration.

Looking to the stars, emerging technologies such as methane-fueled engines and aerospike designs promise exciting possibilities. These advancements are not just incremental improvements; they are potential gateways to new worlds. They hint at a future where Mars is within reach, space travel becomes routine, and the boundaries of human experience expand beyond our wildest dreams.

USAF
BOEING
SOLS
DELTA
STEREO

CHAPTER III
SPACE
INFRASTRUCTURE

EYES IN THE SKY: SATELLITES AND THEIR ORBITS

Satellites are the unsung heroes of modern civilization, silently orbiting Earth and providing critical services that impact our daily lives. From weather forecasting and global communications to navigation and scientific research, satellites play a pivotal role in our understanding and management of the world.

TYPES OF ORBITS: THE CELESTIAL HIGHWAYS OF SPACE TECHNOLOGY

In the vast expanse of space, satellites serve as our eyes, ears, and communication relays, each one carefully placed in a specific orbit to fulfill its unique mission. These orbits are not merely arbitrary paths but meticulously calculated celestial highways, each offering distinct advantages and challenges for the satellites that traverse them.

The choice of orbit is a critical decision in satellite design and deployment, fundamentally shaping a satellite's capabilities, lifespan, and the services it can provide. From weather monitoring and global communications to scientific research and national security, a satellite's orbit is intrinsically linked to its purpose.

In this section, we'll explore the various types of orbits utilized in modern space technology. We'll delve into the characteristics of each orbit, examining how factors such as altitude, inclination, and eccentricity influence a satellite's behavior and functionality. From the bustling low Earth orbit (LEO) teeming with communication satellites and space stations to the distant geostationary orbit (GEO), where weather satellites maintain their vigilant watch, each orbital path offers a unique vantage point for observing and interacting with our planet.
Understanding these orbits is crucial for grasping the current state of space technology and envisioning the future of space exploration and utilization. As we journey through the different types of orbits, we'll uncover the intricate dance of satellites around our planet, each one precisely positioned to extend humanity's reach beyond the confines of Earth.

The primary types of orbits include:

1. Low Earth Orbit (LEO):
- Altitude: 160 to 2,000 kilometers above Earth.
- Characteristics: Short orbital periods (about 90-120 minutes), high resolution for imaging.
- Uses Earth observation, scientific research, and communication constellations like SpaceX's Starlink.

2.   Medium Earth Orbit (MEO):
- Altitude: 2,000 to 35,786 kilometers.
- Characteristics: Longer orbital periods than LEO, typically used for navigation.
- Uses: GPS, GLONASS, and other navigation systems.

3.   Geostationary Orbit (GEO):
- Altitude: Approximately 35,786 kilometers above the equator.
- Characteristics: Satellites appear stationary relative to a point on Earth, providing continuous coverage.
- Uses: Weather monitoring, telecommunications, and broadcast services.

4.   Polar and Sun-Synchronous Orbits (SSO):
- Characteristics: Pass over the poles, allowing global coverage as Earth rotates beneath.
- Uses: Earth observation, environmental monitoring, and reconnaissance.

SATELLITE FUNCTIONS

Satellites serve as the silent workhorses of our modern, interconnected world, performing various critical functions that have become integral to our daily lives. These sophisticated machines, orbiting high above the Earth's surface, act as our eyes in the sky, our voices across vast distances, and our guides in navigation. From enabling global communications and weather forecasting to providing precise positioning data and advancing scientific research, satellites have revolutionized numerous fields and extended human capabilities far beyond our planet's boundaries.

The functions of satellites are as varied as they are vital, each tailored to meet specific needs in science, commerce, security, and exploration. Communication satellites relay television, telephone, and internet signals across the globe, while Earth observation satellites monitor our planet's climate, resources, and potential natural disasters. Navigation satellites form the backbone of GPS systems, enabling precise location services worldwide. Scientific satellites peer into the depths of space, uncovering the mysteries of our universe, while military satellites play crucial roles in national security and intelligence gathering. Satellites serve a variety of functions, each tailored to specific needs:

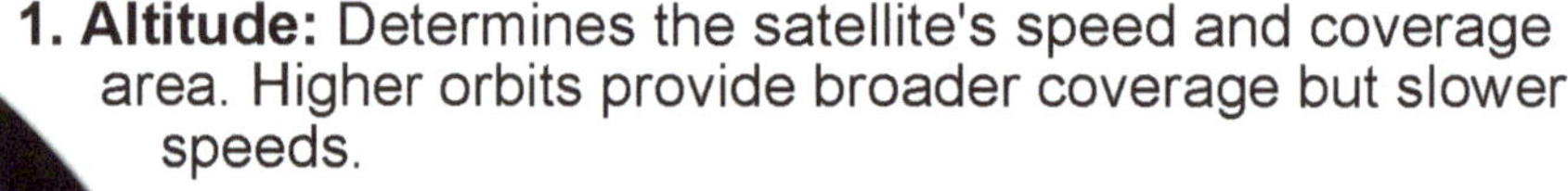

1. Communication Satellites:
• Provide global telecommunication services, including television, radio, internet, and phone.
• Examples: Intelsat, Inmarsat.

2. Weather Satellites:
• Monitor atmospheric conditions, track storms, and provide data for weather forecasting.
• Examples: GOES, Meteosat.

3. Navigation Satellites:
• Enable precise positioning and timing services for navigation and location-based services.
• Examples: GPS, Galileo.

4. Earth Observation Satellites:
• Capture high-resolution images and data for environmental monitoring, agriculture, and urban planning.
• Examples: Landsat, Sentinel.

5. Scientific Satellites:
• Conduct research and gather data on space, Earth, and other celestial bodies.
• Examples: Hubble Space Telescope, James Webb Space Telescope.

Orbital Mechanics

Understanding the mechanics of satellite orbits is crucial for their successful deployment and operation. Key factors include:

1. **Altitude:** Determines the satellite's speed and coverage area. Higher orbits provide broader coverage but slower speeds.

2. **Eccentricity:** Describes the shape of the orbit. Circular orbits have low eccentricity, while elliptical orbits have high eccentricity.

3. **Inclination:** The angle of the orbit relative to Earth's equator. Inclined orbits provide coverage over different latitudes.

Orbital mechanics is a complex field that governs how satellites and other objects move in space. Understanding these principles is crucial for successfully deploying and operating satellites. Some key aspects include:

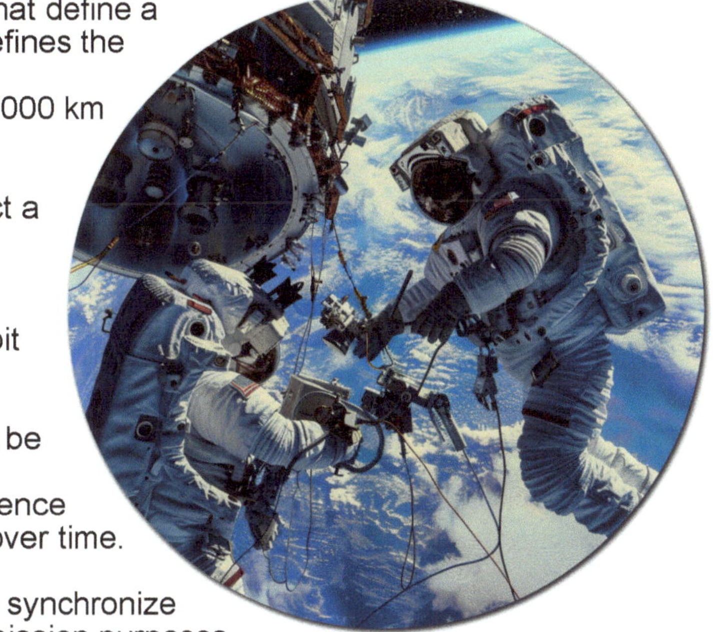

1. Orbital Elements: These are the parameters that define a satellite's orbit, including the semi-major axis: Defines the size of the orbit

2. Types of Orbits: Low Earth Orbit (LEO): 160-2000 km altitude, used for Earth observation and some communication satellites

3. Orbital Perturbations: Various forces can affect a satellite's orbit over time, including Earth's non-spherical gravitational field

4. Orbital Maneuvers: Satellites often need to perform maneuvers to achieve their intended orbit after launch

5. Launch Requirements: The initial velocity and direction needed to achieve a specific orbit must be precisely calculated.

6. Orbital Decay: Satellites in lower orbits experience atmospheric drag, causing their orbits to decay over time. This must be accounted for in mission planning.

7. Synchronization: Some orbits are designed to synchronize with Earth's rotation or other cycles for specific mission purposes.

Understanding these principles allows engineers to design satellite missions, plan orbits, predict satellite positions, and manage the growing population of objects in Earth orbit.

GATEWAY TO THE STARS: ORBITAL TERMINALS

Orbital terminals, commonly known as spaceports, are the critical gateways that facilitate humanity's ventures into space. These facilities are the launchpads for our dreams, enabling the deployment of satellites, the journey of crewed missions, and the cosmos exploration. As the space industry grows, the role of spaceports becomes increasingly vital, supporting a wide range of missions and fostering the development of new technologies and commercial opportunities.

Key Components of Orbital Terminals:

1. Launch Pads:

Function: These are the platforms where rockets are positioned for liftoff. They include essential systems for fueling, electrical connections, and structural support. Modern launch pads feature flame trenches, umbilical towers, and water deluge systems for cooling and protection during launch.

2. Mission Control Centers:

Function: These facilities coordinate and monitor all launch and mission operations aspects. They are equipped with advanced communication systems, real-time data analysis tools, and decision-making support systems. Mission control centers are the nerve centers of space operations.

3. Assembly and Integration Facilities:

Function: These are where rockets and payloads are assembled, integrated, and tested before launch. They include clean rooms to prevent contamination, large assembly bays for rocket stages, and testing equipment to ensure all components function correctly.

4. Recovery and Landing Sites:

Function: These areas are designated for the safe return of reusable rocket stages and crewed spacecraft. They are equipped with tracking systems, retrieval equipment, and infrastructure to support the recovery process.

NOTABLE ORBITAL TERMINALS:

Kennedy Space Center (USA):

Overview: NASA's primary launch center for human spaceflight is located in Florida. It's the historic site of Apollo moon missions and Space Shuttle launches and now supports SpaceX and other commercial launches.

The Kennedy Space Center (KSC) offers several unique features compared to other spaceports:

- Comprehensive infrastructure: KSC manages over 200 facilities spanning over 140,000 acres, providing end-to-end capabilities for integrating, launching, and processing space vehicles and payloads.
- Iconic facilities: KSC houses unique structures like the massive Vehicle Assembly Building, Launch Complex 39 pads, and Orbiter Processing Facilities that are unmatched at other spaceports.
- Multi-user capabilities: KSC supports various government and commercial users, including NASA, SpaceX, United Launch Alliance, and Blue Origin, making it a true multi-user spaceport.

- Full integration capabilities: KSC can uniquely integrate entire space systems in a way no other U.S. site can match, from payload processing to vehicle assembly to launch operations.
- Specialized workforce: Decades of space operations have created a highly skilled local workforce with expertise across all aspects of spaceflight.
- Prime location: Its coastal Florida location provides ideal launch trajectories for a wide range of orbits and missions.
- Historical significance: As the launch site for the Apollo moon missions and Space Shuttle program, KSC has an unparalleled heritage in human spaceflight.
- Research facilities: KSC houses laboratories and test facilities for developing space exploration and utilization technologies.
- Visitor complex: Unlike most spaceports, KSC has a major visitor center providing public education and space exploration outreach.
- Environmental preservation: KSC co-exists with the Merritt Island National Wildlife Refuge, balancing space operations with environmental stewardship.
- Adaptability: KSC has demonstrated the ability to evolve its facilities and operations over time to support new vehicles and mission types, from Apollo to Shuttle to commercial spaceflight.

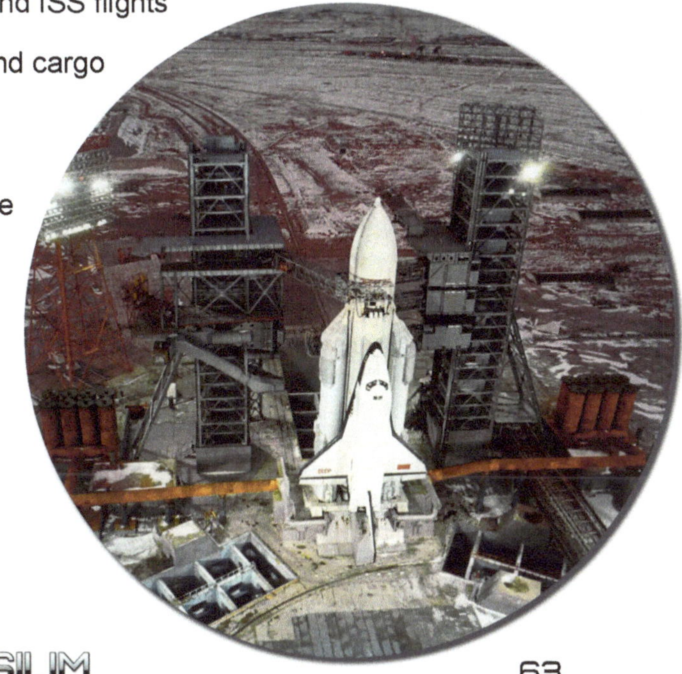

These features collectively make KSC a unique national asset and premier gateway to space that is difficult for other spaceports to replicate.

Baikonur Cosmodrome (Kazakhstan):

Overview: The world's first and largest operational space launch facility. It's the launch site for Soyuz missions to the International Space Station and has been crucial since the early days of space exploration. The Baikonur Cosmodrome supports international space missions in several key ways:

Launch site for International Space Station (ISS) missions:

- All Russian rockets used for manned space missions and ISS flights are launched from Baikonur.
- It has been the primary launch site for sending crews and cargo to the ISS since the station began operations.

Multinational crew launches:
- Baikonur launches Russian cosmonauts, American astronauts, and those from other countries traveling to the ISS.
- After the U.S. Space Shuttle program ended in 2011, Baikonur became the sole launch site for sending crews to the ISS until SpaceX began crewed launches in 2020.

International cooperation:
- The cosmodrome facilitates cooperation in space exploration efforts between Russia, Kazakhstan, the United States, European countries, and others.
- It serves as a hub for international space partnerships, with multiple countries relying on its launch capabilities.

Commercial and scientific satellite launches:
- Baikonur is used to launch satellites for various countries and commercial entities, supporting global communications, Earth observation, and scientific research.
- Infrastructure and facilities:
- The cosmodrome has extensive infrastructure including multiple launch pads, assembly and integration facilities, and mission control centers that support a wide range of international space missions.

Historical significance:
- As the **world's first and largest operational spaceport**, Baikonur carries historical importance that adds prestige to international missions launched from there.

Adaptability:
- Over the years, Baikonur has adapted to support new types of rockets and spacecraft from different countries, demonstrating its flexibility in accommodating international needs.

In summary, the Baikonur Cosmodrome supports international space missions by providing reliable launch capabilities, facilitating multinational crew transport, and serving as a center for global cooperation in space exploration.

Guiana Space Centre (French Guiana):
Overview: The European Space Agency's primary launch site is located near the equator. Its location significantly boosts launch efficiency, especially for geostationary satellites.

Vandenberg Space Force Base (USA):
Overview: This California Space Force base specializes in polar orbit launches and is used for military, scientific, and commercial missions, particularly those requiring polar orbits.

The Future of Orbital Terminals:

Standardization and Flexibility:
- Need: As the number of launch vehicles and missions
increases, there is a growing need for standardized infrastructure that can accommodate various types of rockets and payloads. Standardized interfaces and modular systems can significantly reduce the time and cost associated with preparing different payloads for launch, enhancing overall efficiency. This approach also allows for greater scheduling and resource allocation flexibility, enabling spaceports to handle a higher volume of diverse missions simultaneously.

Public-Private Partnerships:
- Trend: The future of space launch is increasingly commercial, with private sector investment playing a crucial role in spaceport development and operations. Public-private partnerships (PPPs) can leverage the strengths of both sectors, combining government oversight and funding with private innovation and efficiency. These collaborations are essential for developing new technologies, expanding infrastructure, and ensuring that spaceports can meet the growing demands of the global space economy.

Sustainability and Environmental Considerations:

- Challenge: Ensuring spaceport operations are environmentally sustainable and minimizing their impact on local ecosystems is a growing concern. Spaceports are increasingly adopting green technologies, such as biofuels and solar power, to reduce their carbon footprint and environmental impact. Additionally, comprehensive environmental assessments and community engagement are crucial for gaining public support and ensuring spaceport activities do not harm local wildlife and habitats.

Global Expansion:

Trend: New spaceports are being developed worldwide to meet the growing demand for launch services, including locations in Scotland, Norway, and Sweden for small satellite launches. This global expansion is driven by the increasing accessibility of space and the need for more launch capacity to support the burgeoning satellite industry. Emerging spaceports also focus on niche markets, such as polar and sun-synchronous orbits, to provide specialized services that complement existing launch capabilities.

These additional sentences provide a more comprehensive understanding of the future trends and challenges facing orbital terminals, highlighting the importance of standardization, public-private partnerships, sustainability, and global expansion in the evolving space industry.

Orbital terminals are the linchpins of our space endeavors, providing the infrastructure necessary to launch and recover spacecraft. As the space industry continues to expand, these facilities will play an increasingly important role in supporting a diverse array of missions, from satellite deployment to crewed spaceflights.

The evolution of spaceports, driven by standardization, public-private partnerships, sustainability, and global expansion, will ensure that humanity's gateway to the stars remains robust and capable of meeting the challenges of the future. As we continue to push the boundaries of space exploration, the development and enhancement of orbital terminals will be crucial in enabling our journey to the stars.

KEEPING SPACE CLEAN: ORBITAL MAINTENANCE AND DEBRIS MANAGEMENT

As the number of satellites and space missions increases, so does the amount of space debris. Managing this debris is crucial to ensure the sustainability of space activities and the safety of operational satellites. Space debris, also known as space junk, includes defunct satellites, spent rocket stages, and fragments from collisions and explosions. These objects, traveling at high velocities, pose significant risks to active satellites, space stations, and future missions.

Space Debris

Space debris encompasses a wide range of objects, from large defunct satellites to tiny fragments of paint. The real danger lies in the speed at which these objects travel— up to 28,000 kilometers per hour—turning even small pieces into potentially catastrophic projectiles. The sources of space debris are varied, including:

1. Defunct Satellites: Satellites that have reached the end of their operational life but remain in orbit.
2. Rocket Stages: Spent stages of rockets that are discarded during the launch process.
3. Mission-related Objects: Items such as dropped tools, screws, and other equipment lost during space missions.
4. Fragments from Collisions and Explosions: Debris generated from the breakup of satellites and rocket stages due to collisions or explosions.

Risks of Space Debris

The risks posed by space debris are significant and multifaceted:
1. Collisions with Operational Satellites: Even small fragments can cause severe damage to active satellites, potentially leading to loss of functionality and costly repairs or replacements.

2. Threats to Human Spaceflight: Space debris poses a direct threat to crewed missions, including the International Space Station (ISS), which must regularly perform maneuvers to avoid collisions.

3. Increased Costs: Satellite operators incur additional costs for collision avoidance maneuvers and the potential loss of valuable assets.

4. Environmental Impact: Debris from collisions and explosions can create further debris, exacerbating the problem and increasing the risk of a cascading effect known as the Kessler Syndrome.

Debris Management Strategies

Effective management of space debris involves a combination of mitigation and active removal strategies. These strategies are essential to ensure the sustainability of space activities and the safety of operational satellites.

Design Improvements:

- **End-of-Life Disposal:** Satellites and rocket stages are designed with mechanisms to ensure they can be safely deorbited or moved to a graveyard orbit at the end of their operational life. For example, ESA's guidelines require a probability of successful disposal higher than 90%. Companies like Airbus and Thales Alenia Space are incorporating these design improvements into their satellite systems.

- **Passivation:** Removing residual energy sources, such as leftover fuel or batteries, to prevent explosions that could create debris. This involves venting excess propellant and discharging batteries. Lockheed Martin and Northrop Grumman are among the companies that have implemented passivation techniques in their spacecraft designs.

- **Collision Avoidance:** Implementing automated collision avoidance systems to maneuver satellites out of the path of potential collisions. Companies like SpaceX use such systems for their Starlink satellites. Additionally, OneWeb employs similar technologies to ensure the safety of their satellite constellations.

- **Minimizing Intentional Releases:** Reducing the release of mission-related debris, such as lens caps and protective covers, during spacecraft deployment. This can be achieved through tethering or retaining these items until they can be safely disposed of. Companies like Boeing and Surrey Satellite Technology Limited (SSTL) are working on innovative solutions to minimize mission-related debris.

Active Debris Removal (ADR): Technologies and Missions

Active Debris Removal (ADR) involves developing and deploying technologies to capture and remove existing debris from orbit. This is crucial for stabilizing the growth of space debris and preventing future collisions.

Technologies and Missions:

- **Robotic Arms and Nets:** Devices designed to capture and secure debris. For example, ClearSpace, a spin-off from the Swiss EPFL Space Center, is developing the ClearSpace-1 mission to capture and deorbit a defunct satellite using a robotic arm. Airbus is also working on net-based capture systems as part of the RemoveDEBRIS mission.

- **Harpoons and Tethers:** Innovative methods to latch onto debris and drag it into a lower orbit for re-entry. The European Space Agency (ESA) is exploring these technologies as part of its CleanSpace initiative. Northrop Grumman's Mission Extension Vehicle (MEV) also demonstrates the potential for tether-based servicing and debris removal.

- **Laser Systems:** Ground-based or space-based lasers that can nudge debris into lower orbits by imparting small forces. This method is still in the experimental stage but holds promise for future debris management. Companies like Lockheed Martin and the Japanese Aerospace Exploration Agency (JAXA) are researching laser-based debris removal technologies.

Space Traffic Management: Tracking and Monitoring

Space Traffic Management (STM) involves tracking and monitoring space objects to prevent collisions and manage the growing population of satellites and debris.

Tracking and Monitoring:
- **Ground-Based Radars and Telescopes:** These systems track the positions and trajectories of space objects. Companies like LeoLabs provide advanced tracking services using ground-based radar systems. The U.S. Space Surveillance Network (SSN) also plays a crucial role in tracking space debris.
- **Space-Based Sensors:** Satellites equipped with sensors to monitor space debris and provide real-time data. NorthStar Earth & Space Inc. is developing a constellation of satellites for space situational awareness. The European Space Agency (ESA) is also working on space-based debris tracking systems as part of its Space Safety Program.
- **Collision Avoidance Systems:** Automated systems that predict potential collisions and recommend or execute avoidance maneuvers. Share My Space, a French startup, offers collision warning services based on deep learning algorithms. Companies like Lockheed Martin and Boeing are integrating advanced collision avoidance systems into their satellite operations.
- **International Collaboration:** Sharing data and coordinating efforts among spacefaring nations to manage space traffic. The U.S. Space Force's Space Fence radar system and the European Space Agency's Space Debris Office are key players in this effort. The Inter-Agency Space Debris Coordination Committee (IADC) also facilitates international cooperation on space debris management.

Managing space debris is a critical challenge that requires a multifaceted approach, combining mitigation, active removal, and international cooperation. Companies like Astroscale, ClearSpace, OrbitGuardians, and LeoLabs are at the forefront of developing innovative solutions to tackle this issue. As we continue to expand our presence in space, it is essential to address the growing problem of space debris to ensure the long-term sustainability of space activities. By developing and implementing effective debris management strategies, we can protect valuable space assets, ensure the safety of human spaceflight, and preserve the space environment for future generations.

Future Prospects

The future of space debris management looks promising, with advancements in technology and increased international cooperation paving the way for more effective solutions. Key areas of focus include:

1. Enhanced Tracking and Monitoring: Improving the accuracy and coverage of debris tracking systems to better predict and prevent collisions.

2. Innovative Removal Technologies: Developing new methods for capturing and removing debris, such as using ground-based lasers to nudge debris into lower orbits for re-entry.

3. Regulatory Frameworks: Establishing and enforcing international guidelines and agreements to ensure responsible behavior in space and promote the sustainable use of Earth's orbits.

Managing space debris is a critical challenge that requires a multifaceted approach, combining mitigation, active removal, and international cooperation. As we continue to expand our presence in space, it is essential to address the growing problem of space debris to ensure the long-term sustainability of space activities. By developing and implementing effective debris management strategies, we can protect valuable space assets, ensure the safety of human spaceflight, and preserve the space environment for future generations.

Servicing the Satellites: Maintenance in Orbit

In-orbit satellite servicing is an emerging field that promises to extend the lifespan of satellites, reduce costs, and enhance the sustainability of space operations. This innovative approach to satellite maintenance and repair is transforming the space industry by offering solutions to address issues that were previously considered mission-ending.

Types of In-Orbit Servicing:

- **Refueling:** Extends the operational life of satellites by replenishing their fuel supply.
- **Repair and Upgrades:** Performs repairs or upgrades to satellite components, addressing issues that arise during a satellite's lifetime.
- **Relocation:** Moves satellites to new orbits to optimize their performance or avoid collisions.
- **Deorbiting:** Safely removes defunct satellites from orbit to reduce space debris.

The most significant challenges in in-orbit satellite servicing:

Technical Challenges: The primary technical hurdles include precise rendezvous and docking with non-cooperative satellites, especially those not designed for servicing. This requires advanced navigation, control, and docking mechanisms. Robotic manipulation in the zero-gravity environment presents another significant challenge, as even small motions can affect the stability of both the servicing spacecraft and the target satellite. Additionally, the harsh space environment, with extreme temperatures and radiation, complicates servicing operations.

Operational and Economic Challenges: Many servicing tasks need to be performed autonomously due to communication delays, necessitating advanced AI and computer vision systems. The economic viability of servicing missions is a major consideration, as the high cost of these operations must be balanced against the value of extending a satellite's life or capabilities. There's also a risk of creating more space debris if a servicing attempt goes wrong, which could have severe consequences for the space environment and other operational satellites.

Legal and Standardization Challenges: The legal and regulatory landscape for in-orbit servicing is still developing, with questions of liability, ownership, and permission for servicing other entities' satellites yet to be fully addressed. The lack of standardization across different satellite manufacturers complicates servicing operations, as there are no universal interfaces or protocols for attaching to and servicing satellites. This challenge extends to future satellite designs, which will need to incorporate features that facilitate servicing to make these operations more feasible and cost-effective.

Future Prospects and Challenges:

1. Standardization: Developing common interfaces and protocols for servicing operations to ensure compatibility across different satellite manufacturers.

2. Regulatory Framework: Establishing international guidelines and regulations for in-orbit servicing activities to ensure safety and prevent potential conflicts.

3. Debris Mitigation: Using servicing technologies to actively remove space debris and deorbit defunct satellites, contributing to a cleaner space environment.

4. Economic Viability: Balancing the cost of servicing missions against the value of extending satellite lifespans or upgrading capabilities.

5. Technological Advancements: Developing more sophisticated robotic systems, AI-driven autonomy, and advanced propulsion systems to enhance servicing capabilities.

6. Dual-Use Concerns: Addressing potential dual-use nature of servicing technologies, which could have both civilian and military applications.

Major Players in the Industry:

1. Northrop Grumman: Pioneered commercial satellite servicing with their Mission Extension Vehicle (MEV).
2. Astroscale: Japanese company focused on active debris removal and end-of-life services.
3. Maxar Technologies: Developing the OSAM-1 (On-Orbit Servicing, Assembly, and Manufacturing 1) mission in partnership with NASA.
4. SpaceLogistics (a Northrop Grumman subsidiary): Offers Mission Extension Vehicles (MEVs) and develops Mission Robotic Vehicles (MRVs) for more complex servicing tasks.
5. Effective Space Solutions: Developing the Space Drone program for satellite life extension and orbital adjustments.
6. Orbit Fab: Working on creating a network of fuel depots in space, dubbed "Gas Stations in Space."
7. DARPA (Defense Advanced Research Projects Agency): Initiated the Robotic Servicing of Geosynchronous Satellites (RSGS) program.

Conclusion

As we've explored the intricate world of space infrastructure, from the silent sentinels orbiting our planet to the groundbreaking technologies that maintain and service them, one thing becomes abundantly clear: we are at the dawn of a new era in space exploration and utilization.

The satellites that form the backbone of our global communication, navigation, and observation systems are no longer just tools but integral parts of our civilization's infrastructure. As we've seen, the orbits these satellites occupy are carefully chosen celestial highways, each offering unique advantages for specific missions and purposes.

Our journey through the gateways to space - the spaceports and orbital terminals - has revealed the critical role these facilities play in our cosmic endeavors. From the historic launchpads of Kennedy Space Center to the emerging commercial spaceports around the globe, these sites are the springboards for our dreams of space exploration and the engines of a burgeoning space economy.

Perhaps most crucially, we've confronted the growing challenge of space debris and the innovative solutions being developed to address it. The efforts to mitigate, remove, and manage orbital debris underscore our responsibility as stewards of the space environment. Companies like Astroscale, ClearSpace, and Northrop Grumman are pioneering technologies that will be essential for maintaining a sustainable space ecosystem.

The emergence of in-orbit satellite servicing marks a paradigm shift in how we approach space operations. This technology promises to extend the lifespan of valuable space assets, reduce waste, and open up new possibilities for space-based services and exploration.

As we look to the future, the infrastructure we build and maintain in space will be the foundation upon which humanity's cosmic ambitions are realized. From lunar outposts to Martian colonies, from asteroid mining to deep space exploration, every future achievement will rely on the robust and sustainable space infrastructure we are developing today.

The challenges are significant - from the technical hurdles of operating in the harsh environment of space to the regulatory and economic considerations of a rapidly evolving industry. Yet, the potential rewards are even greater. A well-managed space infrastructure could unlock unprecedented scientific discoveries, drive technological innovations, and perhaps even ensure the long-term survival of our species.

As we continue to push the boundaries of what's possible in space, let us remember that every satellite, launch pad, and piece of space technology is a testament to human ingenuity and our unquenchable thirst for knowledge and exploration. The space infrastructure we build today is not just a collection of hardware and facilities—it's the scaffolding upon which we will construct humanity's future among the stars.

CHAPTER IV
LIVING AND WORKING IN SPACE

The dream of living and working in space has captivated humanity for decades, and today, it is becoming a reality through the efforts of space agencies and private companies around the world. The International Space Station (ISS) serves as a unique laboratory where astronauts conduct experiments and test new technologies that will pave the way for future missions to the Moon, Mars, and beyond. Living in space presents numerous challenges, from maintaining life support systems to ensuring the physical and psychological well-being of the crew. As we prepare for longer and more distant missions, understanding and overcoming these challenges is crucial for the success of human space exploration.

In this chapter, we will delve into the critical systems and technologies needed to sustain human life in the harsh environment of space. We will explore the intricacies of Environmental Control and Life Support Systems (ECLSS), which are essential for providing breathable air, clean water, and waste management. Additionally, we will examine the innovative approaches to space farming, which aim to grow food beyond Earth, and the health implications of living in microgravity. From space medicine to logistics and microgravity manufacturing, this chapter will provide a comprehensive overview of the advancements and ongoing research that are making it possible for humans to live and work in space for extended periods.

LIFE SUPPORT: SUSTAINING HUMANS IN SPACE

The harsh environment of space presents unique challenges for human survival. Environmental Control and Life Support Systems (ECLSS) are critical for maintaining a habitable environment for astronauts during long-duration missions. These systems must be highly reliable, efficient, and as self-sustaining as possible to reduce the need for resupply from Earth.

Environmental Control and Life Support Systems (ECLSS):

ECLSS is an integrated network of systems designed to maintain a safe and comfortable environment for the crew. On the International Space Station (ISS), the ECLSS manages atmospheric pressure, fire detection and suppression, oxygen levels, waste management, and water supply. The system is designed to recycle and reuse resources as much as possible to minimize the need for resupply missions.

Air Revitalization and Oxygen Generation:

- **Carbon Dioxide Removal:** The ISS uses systems like the Vozdukh CO2 scrubber and the Carbon Dioxide Removal Assembly (CDRA) to remove CO2 from the air.

- **Oxygen Generation:** The Oxygen Generation System (OGS) on the ISS uses electrolysis to split water molecules into oxygen and hydrogen. The oxygen is released into the cabin atmosphere, while the hydrogen is typically vented into space.

- **Trace Contaminant Control:** Systems like the Trace Contaminant Control System (TCCS) remove harmful contaminants from the air, such as ammonia, methane, and various organic compounds.

Water Recycling and Purification:

The Water Recovery System (WRS) on the ISS recycles wastewater, including urine, hand wash, and condensation, into potable water. This system can recover up to 93% of the water it processes, significantly reducing the need for water resupply from Earth.

The process involves:
- Urine processing through the Urine Processor Assembly (UPA)
- Wastewater processing through the Water Processor Assembly (WPA)
- Multiple filtration and purification steps, including distillation and iodine treatment

Waste Management:
Efficient waste management is crucial for long-duration missions. Current systems include:
- Solid waste collection and storage
- Urine collection and processing for water recovery
- Development of technologies to recycle solid waste into useful materials or fuel

Temperature and Humidity Control:
Maintaining comfortable temperature and humidity levels is essential for crew health and equipment functionality. Systems include:
- Active Thermal Control System (ATCS) to regulate temperature
- Humidity control through condensing heat exchangers
- Air circulation systems to prevent pockets of stagnant air

Challenges of Long-Duration Spaceflight Life Support:
- Reliability and redundancy to ensure continuous operation
- Minimizing power consumption and system mass
- Dealing with microgravity effects on fluid systems
- Radiation protection for both crew and sensitive equipment
- Psychological factors related to recycled resources (e.g., water from processed urine)

Closed-Loop Systems and Bioregenerative Life Support:
Future long-duration missions, such as those to Mars, will require even more efficient closed-loop systems. Research is ongoing in areas such as:
- Bioregenerative life support using plants or algae for air revitalization and food production
- Advanced water recovery systems aiming for near 100% efficiency
- In-situ resource utilization (ISRU) to produce oxygen, water, and fuel from local resources on other planets
- Development of 3D-printed replacement parts to reduce the need for spare part storage

These advanced life support technologies are not only crucial for space exploration but also have potential applications on Earth, particularly in resource-scarce environments or for improving sustainability in urban settings.

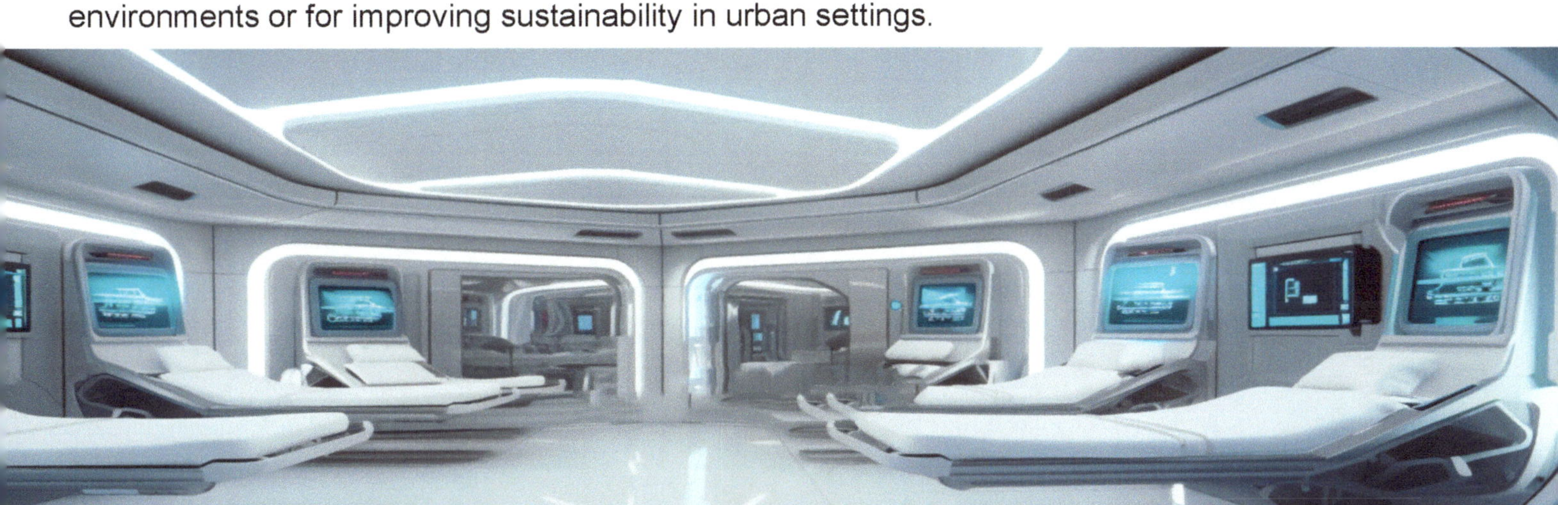

SPACE FARMING: GROWING FOOD BEYOND EARTH

As humanity ventures further into space, the ability to grow food in extraterrestrial environments becomes increasingly critical. Space farming is essential for long-duration missions, such as those to Mars or lunar bases, where resupply from Earth is impractical. This section explores the challenges and technologies involved in cultivating plants in space, highlighting the importance of space agriculture for sustaining human life beyond Earth.

Importance of Space Agriculture for Long-Duration Missions

Space agriculture is vital for providing astronauts with fresh, nutritious food, which is crucial for their health and well-being during extended missions. Fresh produce can supplement pre-packaged meals, offering essential vitamins and minerals that degrade over time in stored food. Additionally, plants play a significant role in life support systems by removing carbon dioxide and producing oxygen, contributing to a self-sustaining environment. Companies like SpaceX and NASA are actively researching and developing these technologies to support future missions to Mars and beyond.

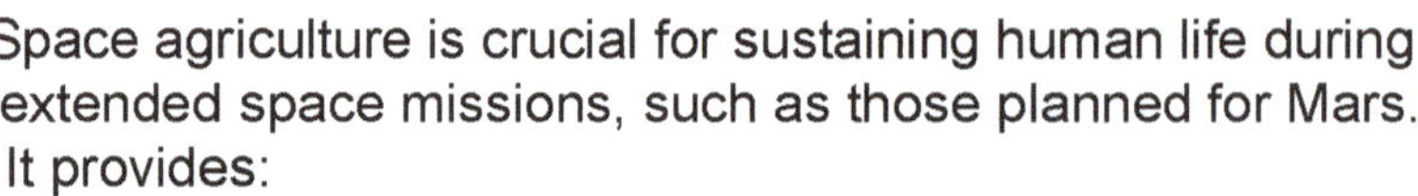

Space agriculture is crucial for sustaining human life during extended space missions, such as those planned for Mars. It provides:
- Fresh, nutritious food to supplement pre-packaged meals, which is essential for maintaining astronaut health and morale.
- Oxygen production and carbon dioxide removal, contributing to a self-sustaining life support system.
- Psychological benefits for crew members, as tending to plants can reduce stress and improve mental well-being.

NASA estimates that a crew of four on a three-year Mars mission would need about 11,000 kg of food. Growing even a portion of this could significantly reduce payload mass and provide vital nutrients.

Challenges of Growing Plants in Microgravity

Microgravity presents unique obstacles for plant growth:
- **Lack of Gravitational Cues:** On Earth, gravity helps orient plant growth, with roots growing downward and stems growing upward. In microgravity, plants lack this directional cue, which can affect their development.
- **Altered Fluid Dynamics:** Water distribution is a significant challenge, as fluids tend to form bubbles in microgravity, potentially drowning roots or leaving them dry.
- **Changes in Plant Physiology:** Microgravity can alter cellular structures and gene expression in plants, potentially affecting growth and nutrient uptake.
- Pathogen Growth: Ensuring adequate air flow and preventing pathogen growth are critical concerns. Research has shown that some plants, like wheat, can grow up to 10% taller in microgravity compared to Earth.

Hydroponics and Aeroponics Systems

To address the challenges of soil-based agriculture in space, scientists have developed hydroponic and aeroponic systems. Hydroponics involves growing plants in a nutrient-rich water solution, while aeroponics suspends plants in the air and mists their roots with nutrient solutions. These soilless methods allow precise control over nutrient delivery and water usage, making them ideal for space environments. Companies like AeroFarms and Plenty are leveraging these technologies for both space and terrestrial applications.

To address the challenges of soil-based agriculture in space, scientists have developed hydroponic and aeroponic systems:

- **Hydroponics:** Plants grow in nutrient-rich water solutions, allowing precise control over nutrient delivery and water usage.
- **Aeroponics:** Roots are suspended in the air and misted with nutrient solutions, reducing water usage and allowing for efficient nutrient absorption.

The XROOTS experiment on the ISS, launched in 2022, is testing these techniques in microgravity. Early results show promise for growing crops like lettuce and tomatoes using these methods. Companies like AeroFarms and Plenty are leveraging these technologies for both space and terrestrial applications.

LED LIGHTING FOR PLANT GROWTH

In the absence of natural sunlight, LED lighting systems provide the necessary light spectrum for photosynthesis. These systems can be tailored to emit specific wavelengths that optimize plant growth and development. The Advanced Plant Habitat (APH) on the ISS, for example, uses a combination of red, blue, green, white, far-red, and infrared LEDs to simulate natural light conditions and support plant health. Companies like Osram and Philips are developing advanced LED systems for space farming.

Specialized LED lighting systems are crucial for plant growth in space:
- Tailored Light Spectra: LED systems can emit specific wavelengths that optimize photosynthesis and plant development.
- Energy Efficiency: LEDs are energy-efficient and can be adjusted to provide the right amount of light for different growth stages.
- Compact and Durable: LED systems are compact and durable, making them ideal for space environments.

The Advanced Plant Habitat on the ISS uses over 180 LEDs to provide precise lighting control. Companies like Osram and Philips are developing advanced LED systems for space farming.

NUTRIENT DELIVERY SYSTEMS

Efficient nutrient delivery is crucial for space farming. Systems like the Veggie plant growth facility on the ISS use "plant pillows" filled with a growth medium and fertilizer to ensure a balanced supply of nutrients. The Passive Orbital Nutrient Delivery System (PONDS) automates this process, reducing the need for manual intervention by astronauts. NASA's Plant Water Management studies are also exploring passive methods to control fluid delivery and uptake in plant systems.

Efficient nutrient delivery is essential in the confined space of a spacecraft:

- Plant Pillows: These are filled with growth media and slow-release fertilizer to ensure a balanced supply of nutrients.
- Passive Orbital Nutrient Delivery System (PONDS): Automates the feeding process, reducing the need for manual intervention by astronauts.
- Controlled Release: Nutrients are released in a controlled manner to match plant growth stages.

NASA's Plant Water Management study is exploring how to optimize water and nutrient delivery in microgravity.

Current Experiments on the ISS

The ISS serves as a test bed for various space farming experiments. The Veggie system, operational since 2014, has successfully grown crops like lettuce, radishes, and zinnias. The Advanced Plant Habitat (APH) is a more sophisticated, enclosed system that requires minimal crew intervention and provides detailed data on plant growth. Recent experiments, such as growing chili peppers, have demonstrated the potential for cultivating a wider variety of crops in space. Companies like SpaceX and Sierra Nevada Corporation are collaborating with NASA on these experiments.

Several ongoing experiments are advancing space agriculture:

- **Veggie:** Operational since 2014, has successfully grown various leafy greens and flowers.
- **Advanced Plant Habitat (APH):** Enclosed, automated system for more complex experiments.
- **Plant Habitat-04:** Successfully grew chile peppers, harvested in 2021.
- **Veg-PONDS-02:** Testing new water and nutrient delivery system.

As of 2024, astronauts have harvested and consumed over 50 crops grown on the ISS. Companies like SpaceX and Sierra Nevada Corporation are collaborating with NASA on these experiments.

FUTURE PROSPECTS FOR LUNAR AND MARTIAN AGRICULTURE

Looking ahead, space agencies and private companies are exploring the feasibility of growing food on the Moon and Mars. Research is focused on using local resources, such as lunar regolith, as a growth medium and developing bioregenerative life support systems that integrate plant cultivation with waste recycling and air revitalization. The goal is to create self-sustaining habitats that can support human life for extended periods. Companies like Interstellar Lab and Blue Origin are at the forefront of developing these technologies. NASA's Artemis program aims to establish a sustainable lunar presence by 2028, including capabilities for food production. Companies like Interstellar Lab and Blue Origin are at the forefront of developing these technologies.

Space farming is a critical component of future space exploration, offering solutions to the challenges of long-duration missions and the potential for self-sustaining habitats on other planets. By overcoming the unique challenges of growing plants in microgravity and developing advanced technologies for nutrient delivery and lighting, scientists are paving the way for sustainable human presence beyond Earth.

ADAPTING TO ZERO-G: HEALTH IN MICROGRAVITY

The absence of gravity in space, known as microgravity, presents unique challenges to the human body. Astronauts must adapt to this environment, which can have both short-term and long-term effects on their health. Understanding these effects is crucial for ensuring the well-being of space travelers and developing effective countermeasures.

Immediate Effects of Microgravity

1. Fluid Shift: Upon entering microgravity, bodily fluids redistribute towards the upper body, causing Facial puffiness and congestion.

2. Space Motion Sickness: Many astronauts experience disorientation and nausea during the first few days in space due to conflicting sensory inputs in the absence of gravity.

3. Balance and Coordination: The vestibular system, responsible for balance, is significantly affected. Astronauts often experience difficulty with spatial orientation and hand-eye coordination.

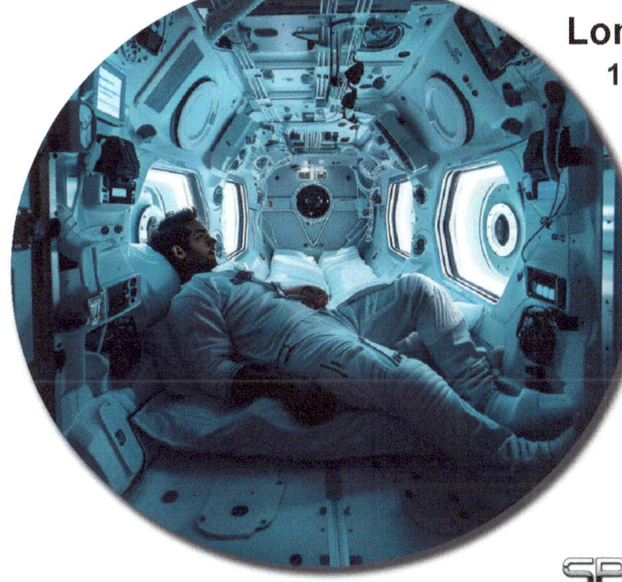

Long-Term Health Impacts

1. Musculoskeletal System: Muscle Atrophy- Without the constant fight against gravity, muscles begin to weaken and shrink.

2. Cardiovascular System: Decreased blood volume

3. Vision and Ocular Health: Spaceflight Associated Neuro-ocular Syndrome (SANS): Changes in intracranial pressure can lead to vision problems and structural changes in the eye.

4. Immune System: Altered immune function, potentially increasing susceptibility to illness

5. Radiation Exposure: Increased risk of cancer and other health issues due to higher radiation levels in space.

Countermeasures and Adaptations

1. Exercise Regimens: Astronauts on the ISS exercise for about 2 hours daily using specialized equipment to combat muscle and bone loss.
2. Nutrition and Supplementation: Carefully planned diets and supplements to support bone health and overall nutrition.
3. Artificial Gravity Concepts: Research into rotating habitats or short-arm centrifuges to provide periods of artificial gravity.
4. Pharmacological Interventions: Medications to prevent bone loss, such as bisphosphonates.
5. Specialized Garments: Compression garments to assist with fluid redistribution.
6. Psychological Support: Regular communication with ground control and family

Ongoing Research

Scientists continue to study the effects of microgravity on the human body, using both space\-based experiments and Earth-based analogs like bed rest studies. Key areas of focus include:

- **Genetic and molecular changes in microgravity**
- **Long-term effects of space radiation**
- **Development of more effective countermeasures for extended missions**
- **Understanding individual variability in adaptation to microgravity**

As we plan for longer missions and potential colonization of other planets, adapting to and mitigating the health effects of microgravity remains a critical challenge in space exploration.

Companies and Organizations Involved in Microgravity Health Research

1. NASA: Leads extensive research through its Human Research Program (HRP)
2. SpaceX: Collaborates with NASA on ISS resupply missions, carrying research payloads
3. Blue Origin: New Shepard suborbital vehicle offers brief periods of microgravity for research
4. Boeing: Starliner spacecraft designed to transport astronauts to the ISS
5. Axiom Space: Developing commercial modules for the ISS and a future private space station
6. Sierra Space: Developing the Dream Chaser spaceplane for cargo and crew transport
7. Northrop Grumman: Provides Cygnus cargo spacecraft for ISS resupply missions
8. Space Tango: Specializes in automated microgravity research and manufacturing platforms
9. Made In Space: Focuses on 3D printing and manufacturing technologies for use in microgravity
10. Techshot: Creates hardware for microgravity research, including bioprinters and cell culturing devices

The journey to understanding and mitigating the effects of microgravity on human health is a testament to human ingenuity and resilience. As we have explored, the absence of gravity presents a myriad of challenges, from fluid shifts and muscle atrophy to bone density loss and vision changes. These physiological adaptations, while significant, are not insurmountable. Through rigorous research and innovative countermeasures, we are making strides in ensuring the health and well-being of astronauts during long-duration space missions.

Organizations like NASA, SpaceX, Blue Origin, and others are at the forefront of this research, developing technologies and strategies to combat the adverse effects of microgravity. NASA's Human Research Program (HRP) and the extensive studies conducted on the International Space Station (ISS) provide invaluable data that drive the development of effective countermeasures.

The implications of this research extend beyond space exploration. The insights gained from studying human physiology in microgravity have the potential to revolutionize medical science on Earth. From developing new treatments for osteoporosis and muscle atrophy to understanding the mechanisms of aging and immune system function, the benefits of space medicine are far-reaching. As we continue to push the boundaries of human exploration, the knowledge and technologies developed to adapt to zero gravity will play a pivotal role in ensuring the success and sustainability of our ventures into the cosmos.

SPACE MEDICINE: ENSURING ASTRONAUT HEALTH

Introduction to Space Medicine

Space medicine is a specialized field dedicated to understanding and mitigating the health risks associated with space travel. It encompasses the study, prevention, and treatment of the physiological and psychological effects of spaceflight on the human body. As we prepare for long-duration missions to the Moon, Mars, and beyond, ensuring astronaut health becomes increasingly critical. The scope of space medicine extends from pre-flight medical screening and preparation to in-flight health monitoring and post-flight rehabilitation.

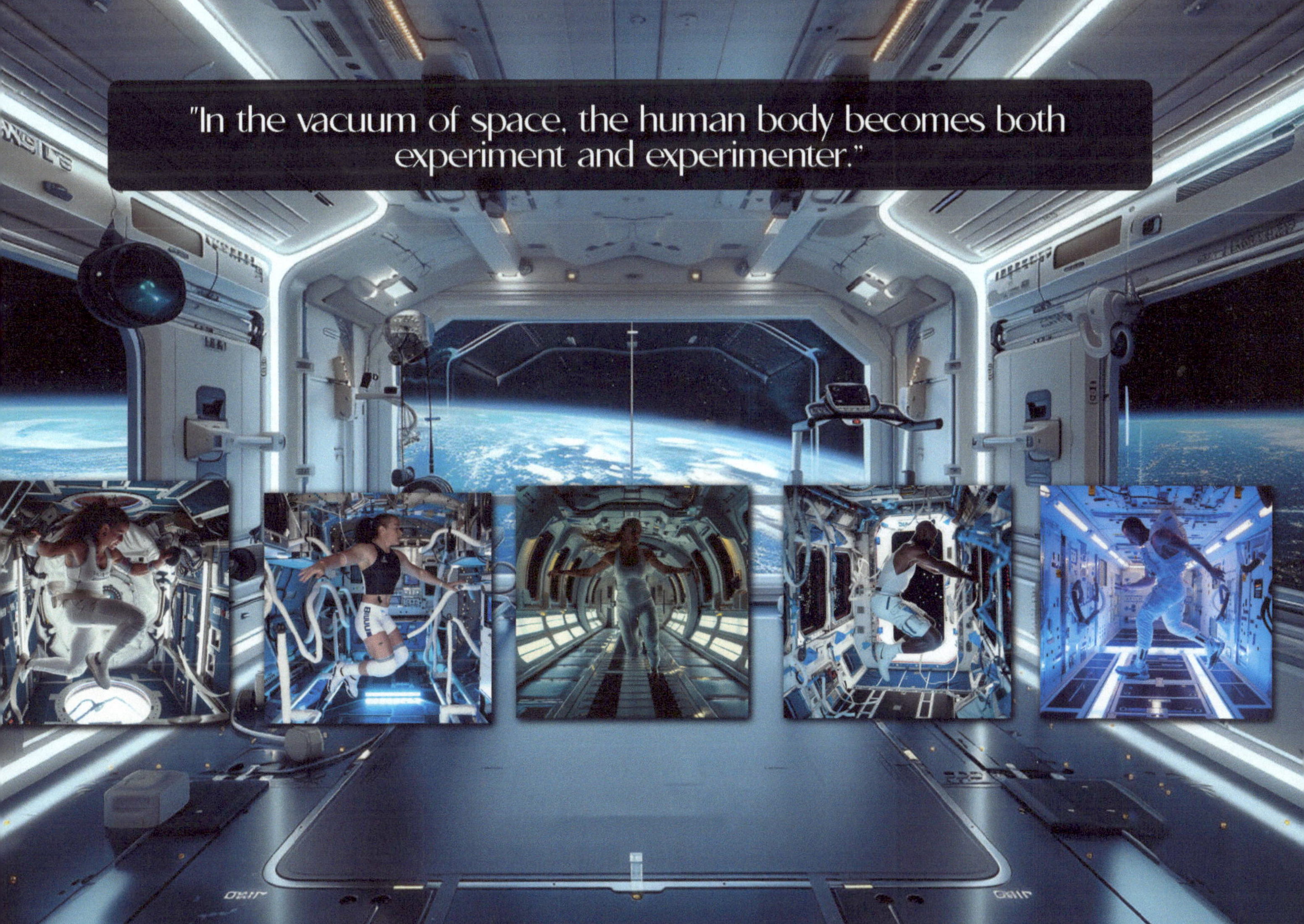

Physiological Challenges in Space

The space environment presents several significant physiological challenges:

• **Bone and Muscle Loss:** In microgravity, astronauts can lose up to 1-1.5% of their bone mass per month, primarily in weight-bearing bones. Muscle atrophy also occurs rapidly, with some studies showing up to a 20% loss in muscle volume on long-duration missions.

• **Cardiovascular Deconditioning:** The absence of gravity leads to fluid shifts in the body, affecting cardiovascular function. This can result in decreased plasma volume, cardiac atrophy, and orthostatic intolerance upon return to Earth.

• **Vision Changes (Spaceflight Associated Neuro-ocular Syndrome - SANS):** Approximately 70% of astronauts on long-duration missions experience some degree of optic disc edema, potentially due to increased intracranial pressure in microgravity.

• **Immune System Alterations:** Studies have shown that spaceflight can suppress immune function, potentially increasing susceptibility to infections and reactivation of latent viruses.

• **Radiation Exposure:** Beyond low Earth orbit, astronauts are exposed to higher levels of cosmic radiation, which can increase the risk of cancer and other health issues.

Psychological Health in Space

The psychological challenges of spaceflight are equally important:

• **Isolation and Confinement:** Extended periods in the confined space of a spacecraft, far from Earth, can lead to feelings of isolation and psychological stress.

• **Stress Management:** Astronauts must manage high-stress situations while maintaining peak performance, requiring robust mental health support systems.

• **Sleep Disorders:** Disruption of circadian rhythms due to the absence of normal day-night cycles can lead to sleep disturbances, affecting both physical and mental health.

Medical Technologies for Space

Several advanced medical technologies are being developed to address the unique challenges of space medicine:

• **Telemedicine and Remote Health Monitoring:** Telemedicine allows astronauts to consult with medical professionals on Earth in real time. Remote health monitoring systems track vital signs and other health metrics, enabling early detection of potential issues.

• **Portable Diagnostic Tools:** Devices like portable ultrasound machines and handheld diagnostic tools are essential for in-flight medical assessments.

• **3D Printing of Medical Supplies:** 3D printing technology can produce medical tools and even bioprinted tissues on-demand, reducing the need for extensive medical supplies.

Countermeasures and Treatments

To mitigate the adverse effects of spaceflight, several countermeasures and treatments are employed:

• **Exercise Regimens and Equipment:** Astronauts on the ISS exercise for about 2 hours daily using specialized equipment like treadmills, stationary bikes, and resistance machines to combat muscle and bone loss.

• **Pharmacological Interventions:** Medications such as bisphosphonates are used to prevent bone loss. Other drugs may be used to manage sleep disorders and other health issues.

• **Artificial Gravity Concepts:** Research into rotating habitats or short-arm centrifuges aims to provide periods of artificial gravity to mitigate the effects of microgravity.

Emergency Medical Procedures in Space

Emergency medical procedures in space require extensive training and preparation:

- **Training and Preparation:** Astronauts undergo rigorous medical training to handle emergencies, including basic surgical procedures and advanced life support.
- **Limitations and Challenges:** The confined space and limited medical supplies on spacecraft pose significant challenges for emergency medical care.

Earth Benefits of Space Medicine Research

Research in space medicine has numerous benefits for healthcare on Earth:

- Spin-off Technologies: Innovations developed for space medicine, such as telemedicine and portable diagnostic tools, have applications in remote and under-served areas on Earth.
- Advancements in Telemedicine and Remote Healthcare: The technologies and practices developed for space missions are improving telemedicine and remote healthcare delivery on Earth.

Future Directions in Space Medicine

As we prepare for long-duration missions to Mars and beyond, space medicine will continue to evolve:

- Preparing for Long-Duration Missions to Mars: Research is focused on developing more effective countermeasures and medical technologies to support astronauts on multi-year missions.
- Potential for Off-Earth Medical Facilities: Future space habitats may include dedicated medical facilities to provide comprehensive healthcare for astronauts.

Companies Leading the Way in Space Life Support Systems

Several companies are at the forefront of developing space life support systems:

- Honeywell: A key player in environmental control and life support systems (ECLSS) for decades, from the Apollo missions to the ISS. They are developing next-generation systems for deep space exploration.
- Collins Aerospace provides critical ECLSS components and systems, including for the ISS. It is also developing advanced technologies for future long-duration missions.
- Paragon Space Development Corporation specializes in life support and thermal control systems for extreme environments. It has developed systems for the ISS and is developing technologies for lunar and Mars missions.

- Boeing: Involved in life support systems for the ISS and developing technologies for future space stations and deep space habitats.
- Northrop Grumman: Provides life support components for the ISS and is developing systems for NASA's Artemis program and future space stations.
- SpaceX: Developing life support systems for their Crew Dragon spacecraft and future Mars missions.
- Blue Origin: Working on life support technologies for their New Shepard vehicle and proposed Orbital Reef space station.
- Sierra Space: Developing life support systems as part of their Dream Chaser spaceplane and proposed commercial space station projects.
- Axiom Space: Designing life support systems for their planned commercial space station modules.
- Lockheed Martin: Involved in life support system development for NASA's Orion spacecraft and other deep space exploration projects.

Opportunities for STEM Students

STEM students interested in space medicine and life support sciences have several pathways to enter the field:

- NASA's Space Life Sciences Training Program (SLSTP): Offers undergraduate and graduate students hands-on experience in space life science disciplines.

- NASA STEM Engagement Program: This program provides various opportunities, including internships and research projects related to space medicine and life support systems.
- National Space Grant College and Fellowship Project: Offers valuable learning experiences for students interested in space-related fields.
- Internships and Entry-Level Positions: Many private companies involved in space medicine research, such as SpaceX, Blue Origin, and Axiom Space, offer internships and entry-level positions for STEM graduates.
- University Research Partnerships: Universities with strong aerospace and biomedical engineering programs often have research partnerships with NASA and private space companies, providing students with opportunities to engage in cutting-edge research.

Space medicine is a critical field for the future of space exploration, addressing the unique medical challenges of space environments and developing innovative technologies and practices to ensure astronaut health. Through collaborations between NASA, private companies, and research institutions, significant progress is being made in understanding and mitigating the physiological and psychological effects of spaceflight. As we prepare for long-duration missions to Mars and beyond, the advancements in space medicine will play a crucial role in ensuring the success and sustainability of human space exploration.

Future Prospects:

The increasing commercialization of low Earth orbit and plans for lunar bases will create new opportunities for research and development in this field. STEM students entering this area can expect to work on challenges ranging from developing closed-loop life support systems to creating new medical technologies for use in space environments.

LOGISTICS IN SPACE: MOVING AND MANAGING RESOURCES

As humanity ventures further into space, the logistics of moving and managing resources become increasingly complex and critical. Space logistics encompasses the planning, execution, and management of transporting humans, equipment, and supplies to, from, and within space. This section explores the unique challenges and innovative solutions in space logistics, highlighting the importance of efficient resource management for the success of space missions.

Introduction to Space Logistics

Space logistics is the science of planning and carrying out the movement of humans and materials to, from, and within space. It involves the design, development, acquisition, storage, movement, distribution, maintenance, and disposal of space materials. Additionally, it includes the movement, evacuation, and hospitalization of people in space, as well as the construction, maintenance, and operation of facilities to support human and robotic space operations.

Categories of Space Logistic Services

Space logistics services can be broadly categorized into five main areas:

- Access to Space: Launch services that transport payloads and crew from Earth to space.
- Space Logistics for Crewed Space Stations: Managing the supply chain for space stations, including resupply missions, waste management, and maintenance.
- Space Logistics for Payloads: Transporting and managing scientific instruments, experiments, and other payloads in space.
- Space Logistics for Spacecraft: This includes providing services such as refueling, repair, and maintenance for satellites and other spacecraft.
- Space Situational Awareness: Monitoring and managing space traffic to avoid collisions and ensure safe operations.

Challenges of Space Logistics

Space logistics presents unique challenges that differ significantly from terrestrial logistics:

- Distance and Time: The vast distances and time required for space travel introduce significant logistical challenges. For example, a mission to Mars can take several months, requiring careful planning and resource management.
- Microgravity and Vacuum: The microgravity environment and vacuum of space affect the handling and storage of materials, requiring specialized equipment and procedures.
- Radiation: Spacecraft and cargo must be protected from cosmic radiation, which can damage equipment and pose health risks to astronauts.
- Cost: Launching materials into space is expensive, necessitating efficient use of space and weight limitations.

Key Players in Space Logistics

Several companies and organizations are leading the way in space logistics:

- NASA: Manages the logistics for the International Space Station (ISS) and is developing systems for future lunar and Mars missions.
- SpaceX: Provides launch services and is developing the Starship for long-duration missions. SpaceX also collaborates with NASA on ISS resupply missions.
- Northrop Grumman: Offers in-orbit satellite servicing through its SpaceLogistics subsidiary, including the Mission Extension Vehicle (MEV) for satellite life extension.
- Blue Origin: Developing the New Glenn rocket and the Orbital Reef space station, which will require robust logistics support.
- Sierra Space: Working on the Dream Chaser spaceplane for cargo and crew transport, as well as commercial space station projects.
- Axiom Space: Planning commercial modules for the ISS and a future private space station, focusing on logistics and resource management.

Innovative Solutions in Space Logistics

Innovative technologies and approaches are being developed to address the challenges of space logistics:

- Reusable Rockets: Companies like SpaceX and Blue Origin are developing reusable rockets to reduce launch costs and increase the frequency of space missions.
- In-Orbit Servicing: Northrop Grumman's MEV and other in-orbit servicing technologies enable the repair, refueling, and maintenance of satellites, extending their operational life.
- 3D Printing: Additive manufacturing in space allows for the on-demand production of tools, spare parts, and even habitats, reducing the need for extensive cargo shipments.
- Autonomous Systems: Autonomous spacecraft and robotic systems can perform complex logistics tasks, such as docking, cargo transfer, and maintenance, with minimal human intervention.
- Space Depots: Orbital fuel depots and logistics hubs, like those proposed by Orbit Fab and Gateway Galactic, provide refueling and resupply services for spacecraft, enabling longer missions and reducing the need for frequent returns to Earth.

Future Directions in Space Logistics

The future of space logistics is promising, with numerous opportunities and challenges on the horizon:

- Lunar and Martian Logistics: As missions to the Moon and Mars become more frequent, efficient logistics systems will be essential for transporting materials, equipment, and personnel to and from these celestial bodies.
- Space-Based Manufacturing and Mining: The potential to utilize resources from the Moon, asteroids, and other celestial bodies for in-space manufacturing and mining will require advanced logistics systems to transport and process these materials.
- Space Tourism and Commercialization: The growth of space tourism and commercial space activities will drive the need for comprehensive logistics solutions to support a wide range of services, from transportation to hospitality. As we look to the future, the development of robust logistics systems will be essential for exploring and utilizing the vast resources of space, ensuring the success of missions to the Moon, Mars, and beyond.

MICROGRAVITY MANUFACTURING: THE FUTURE OF IN-SPACE PRODUCTION

Microgravity manufacturing represents a revolutionary frontier in space exploration and industrial production. The unique environment of space, characterized by near-zero gravity, offers unprecedented opportunities for creating materials and products that are difficult or impossible to produce on Earth. As space technology advances and access to orbit becomes more affordable, the potential for in-space manufacturing is rapidly expanding.

Advantages of Microgravity Manufacturing

1. Absence of Convection and Sedimentation: In microgravity, the lack of convection currents and sedimentation allows for more uniform material mixing and the creation of purer, more homogeneous substances.
2. Containerless Processing: The ability to manipulate materials without containers eliminates contamination and creates ultra-pure materials.
3. Perfect Spheres and Uniform Crystals: Without gravity's influence, liquids naturally form into perfect spheres, and crystals can grow larger and more uniformly.
4. Enhanced Protein Crystallization: Microgravity enables the growth of larger, more perfect protein crystals, crucial for pharmaceutical research and drug development.
5. Unique Alloys and Composites: The space environment allows for the creation of alloys and composite materials with properties unattainable on Earth.

Key Areas of Development

1. Semiconductor Manufacturing: Companies are exploring the potential for producing higher-quality semiconductors in space, with reduced defects and improved performance.
2. Fiber Optics: ZBLAN, a type of exotic optical fiber, shows promise when manufactured in microgravity, potentially offering significantly lower signal loss over long distances.
3. Bioprinting and Tissue Engineering: Microgravity allows for the creation of more complex 3D tissue structures, potentially revolutionizing organ transplantation and drug testing.
4. Advanced Materials: Metal foams, superalloys, and other advanced materials could be produced with unique properties beneficial for various industries on Earth.
5. Pharmaceutical Research: The ability to grow larger, more perfect protein crystals in space could accelerate drug discovery and development processes.

Current Initiatives and Future Prospects

1. International Space Station (ISS) Experiments: The ISS serves as a test bed for various microgravity manufacturing experiments, including 3D printing, crystal growth, and materials science research.
2. Commercial Ventures: Companies like Made In Space, Space Forge, and Redwire Space are developing technologies for in-space manufacturing and exploring commercial applications.
3. Reusable Manufacturing Platforms: Development of reusable satellites and spacecraft designed specifically for manufacturing and returning products to Earth.
4. Lunar and Martian Manufacturing: Future plans for lunar and Martian bases include in-situ resource utilization (ISRU) and manufacturing capabilities to support long-term human presence.

Benefits of In-Space Manufacturing

In-space manufacturing benefits from microgravity environments in several key ways:

1. Improved material properties:
• Microgravity allows for more uniform mixing of materials, creating unique alloys and compositions that are difficult or impossible to produce on Earth.
• Crystals can grow larger and more perfectly in microgravity, which is beneficial for semiconductors, protein crystals for drug development, and other applications.
• The absence of sedimentation and buoyancy allows for more precise control over material formation.

2. Enhanced manufacturing processes:
• Lack of convection currents provides a quiescent environment that can minimize defects in materials.
• Surface tension effects dominate in microgravity, allowing for more uniform and precise layering in thin film deposition.
• Containerless processing is possible, eliminating contamination from container walls.

3. Unique product capabilities:
• Fiber optic cables (like ZBLAN) can be produced with fewer defects, potentially offering significantly lower signal loss.
• Metal alloys can be created with more even distribution of elements, improving their properties.
• Biological tissues and organs can be grown in 3D without the need for scaffolding, as gravity doesn't cause them to collapse.

4. New possibilities for pharmaceuticals:
• Protein crystals grown in space can be larger and more uniform, aiding in drug discovery and development.
• Microgravity allows for unique drug formulations that may have improved bioavailability or extended shelf-life.

5. Advanced materials research:
• The space environment allows for studying material behaviors and creating new materials that are impossible on Earth due to gravitational interference.

6. Semiconductor manufacturing:
• The vacuum of space and lack of contaminants could potentially lead to higher-quality semiconductors with fewer defects.

7. Biomedical applications:
• Microgravity enables the creation of more complex 3D tissue structures, which could revolutionize organ transplantation.

While these benefits are significant, challenges remain, including the high cost of space launches, the need for specialized equipment, and the complexities of operating in the space environment. However, as launch costs decrease and technology advances, in-space manufacturing becomes an increasingly viable option for creating unique, high-value products.

Challenges and Considerations in Space Manufacturing

Cost: Despite significant reductions in launch costs (from $65,000 per kilogram to $1,500 per kilogram in low Earth orbit), space manufacturing remains expensive. The high costs are associated with:
• Specialized equipment development for microgravity environments
• Transportation of raw materials to space
• Maintenance of orbital facilities
• Crew training and support

For widespread adoption, companies must demonstrate a clear return on investment. This may be achievable for high-value, low-volume products like specialized pharmaceuticals or advanced materials that benefit from microgravity conditions. However, for most products, the cost-benefit analysis remains challenging.

Scale: Current manufacturing capabilities in space are extremely limited. The International Space Station (ISS) has some small-scale manufacturing experiments, but nothing close to commercial production levels. Challenges include:

- Limited physical space in current orbital facilities
- Power constraints
- Lack of large-scale automated manufacturing systems designed for space

Significant expansion is needed to achieve commercial viability. This may require dedicated manufacturing satellites or space stations, which represent massive investments.

Quality Control: Ensuring consistent product quality in the space environment presents unique challenges:

- Microgravity affects material behavior and manufacturing processes in ways that are not fully understood
- Radiation exposure can impact material properties and electronic systems
- Limited ability for human intervention in automated processes
- Difficulty in implementing traditional quality control measures in space

Developing new quality assurance methods suitable for space environments is crucial. This may involve advanced sensors, real-time monitoring systems, and AI-driven quality control processes.

Regulatory Framework:
The regulatory landscape for space-manufactured products is still evolving. Key issues include:

- Determining which nation's laws apply to products manufactured in orbit
- Establishing safety and quality standards for space-manufactured goods
- Addressing potential environmental concerns related to space manufacturing
- Developing protocols for the disposal of manufacturing waste in space

International cooperation will be essential to create a comprehensive regulatory framework that encourages innovation while ensuring safety and environmental protection.

Return Logistics:
Safely and cost-effectively returning manufactured goods to Earth is a critical challenge:

- Re-entry vehicles must protect delicate products from extreme heat and g-forces
- The volume of returnable cargo is currently very limited
- Frequent return missions could significantly increase overall costs

Innovations in re-entry vehicle design and the development of more efficient return logistics systems are needed. Some companies, like SpaceX with its Starship, are working on larger, more cost-effective return capabilities.

Additional Considerations

Workforce Development:
Training a specialized workforce for space manufacturing presents unique challenges. This includes:
- Developing new skill sets for engineers and technicians
- Creating training programs for astronauts or robotic operators
- Addressing the psychological challenges of long-duration space missions for manufacturing crews

Supply Chain Management:
Managing a supply chain that extends into orbit adds complexity:
- Coordinating timely delivery of raw materials to space facilities
- Ensuring redundancy in critical components to mitigate risks
- Developing new inventory management systems for space environments

Energy Requirements: Manufacturing processes often require significant energy. In space, this presents challenges:
- Developing more efficient solar power systems
- Exploring alternative energy sources like nuclear power for space applications
- Balancing energy needs between manufacturing and other critical systems

Market Development: Creating demand for space-manufactured products is crucial. This involves:
- Identifying products that truly benefit from space manufacturing
- Educating potential customers about the unique properties of space-made goods
- Developing new markets that may not exist for Earth-manufactured products

While space manufacturing offers exciting possibilities, overcoming these challenges requires significant technological advancements, substantial investment, and innovative problem-solving. As costs continue to decrease and capabilities improve, we may see niche applications become economically viable first, paving the way for broader adoption in the future.

The unique properties of materials produced in microgravity could lead to breakthroughs in electronics, medicine, and advanced materials, potentially transforming numerous sectors on Earth. While challenges remain, the future of in-space production looks promising, with the potential to create a new era of innovation and economic opportunity both in orbit and on our home planet.

Leading Companies in In-Space Manufacturing

1. Made In Space (now part of Redwire Space)
   - Focus: 3D printing and additive manufacturing in space.
   - Notable Projects: First 3D printer on the ISS, Archinaut One (a spacecraft capable of manufacturing and assembling large structures in space).
   - Website: [Redwire Space] (https://redwirespace.com/)

2. Varda Space Industries
   - Focus: Manufacturing advanced materials in space and returning them to Earth.
   - Notable Projects: Developing re-entry capsules to bring manufactured goods back to Earth.
   - Website: [Varda Space Industries] (https://varda.com/)

3. Space Forge
   - Focus: Manufacturing high-purity metals and advanced materials in space.
   - Notable Projects: ForgeStar satellite for customizable in-space manufacturing.
   - Website: [Space Forge] (https://spaceforge.com/)

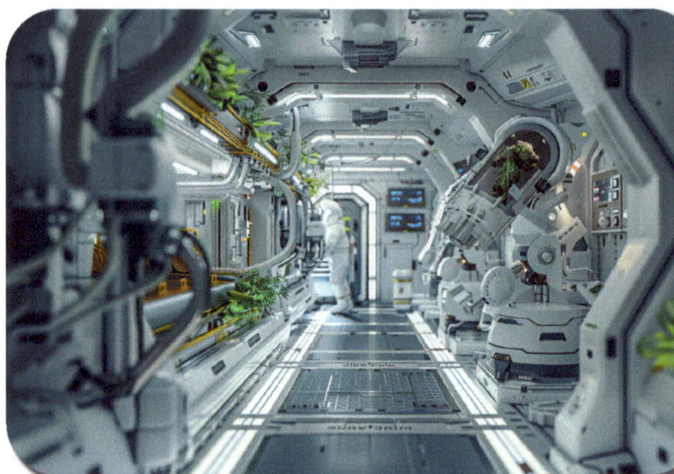

4. LambdaVision
   - Focus: Manufacturing protein-based retinal implants in microgravity.
   - Notable Projects: Developing artificial retinas to restore vision in patients with retinal degeneration.
   - Website: [LambdaVision](https://lambdavision.com/)

5. Axiom Space
   - Focus: Developing commercial space station modules with manufacturing capabilities.
   - Notable Projects: Building modules that will attach to the ISS and eventually form a standalone commercial space station.
   - Website: [Axiom Space] (https://axiomspace.com/)

6. Sierra Space
   - Focus: Developing the Dream Chaser spaceplane and commercial space habitats.

Notable Projects include Dream Chaser for cargo and crew transport and Large Integrated Flexible Environment (LIFE) habitat.
   - Website: [Sierra Space] (https://sierraspace.com/)

7. Northrop Grumman
   - Focus: In-orbit servicing and manufacturing.
   - Notable Projects: Mission Extension Vehicle (MEV) for satellite life extension, developing Persistent Platform for autonomous and robotic capabilities.
   - Website: [Northrop Grumman] (https://northropgrumman.com/)

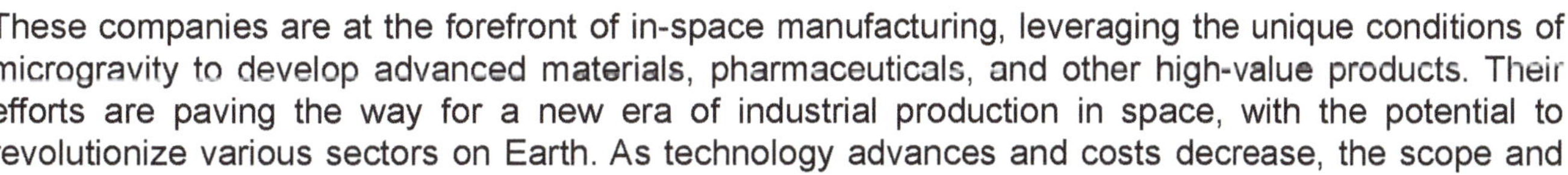

8. Blue Origin
   - Focus: Developing space transportation and infrastructure for in-space manufacturing.
   - Notable Projects: New Glenn rocket, Orbital Reef space station.
   - Website: [Blue Origin] (https://blueorigin.com/)
9. SpaceX
   - Focus: Space transportation and infrastructure development.
   - Notable Projects: Starship for long-duration missions, Dragon spacecraft for ISS resupply and crew transport.
   - Website: [SpaceX](https://spacex.com/)
10. Redwire Space
   - Focus: Space infrastructure and manufacturing.
   - Notable Projects: Developing technologies for in-space manufacturing, including bioprinting and advanced materials.
   - Website: [Redwire Space] (https://redwirespace.com/)
11. Allevi Inc.
   - Focus: 3D bioprinting for healthcare applications.
   - Notable Projects: Developing bioprinters and bioinks to create tissues and organs in space.
   - Website: [Allevi Inc.] (https://allevi3d.com/)
12. Global Graphene Group, Inc.
   - Focus: Advanced materials manufacturing, including graphene and solid-state batteries.
   - Notable Projects: Developing high-performance materials for use in space and on Earth.
   - Website: [Global Graphene Group] (https://theglobalgraphenegroup.com/)
13. SpacePharma
   - Focus: Microgravity research platforms for pharmaceutical and biotech applications.
   - Notable Projects: Providing small labs for space-based biology, chemistry, and materials science research.
   - Website: [SpacePharma](https://spacepharma.com/)
14. ThinkOrbital
   - Focus: Large-scale orbital platforms for in-space research and manufacturing.
   - Notable Projects: ThinkPlatforms and CONTESA (Construction Technologies for Space Applications).
   - Website: [ThinkOrbital](https://thinkorbital.com/)
15. Vast
   - Focus: Developing microgravity and artificial gravity space stations.
   - Notable Projects: Haven-1 is a commercial destination for research and manufacturing.
   - Website: [Vast](https://vast.space/)

These companies are at the forefront of in-space manufacturing, leveraging the unique conditions of microgravity to develop advanced materials, pharmaceuticals, and other high-value products. Their efforts are paving the way for a new era of industrial production in space, with the potential to revolutionize various sectors on Earth. As technology advances and costs decrease, the scope and scale of in-space manufacturing are expected to grow, driving innovation and economic growth both in orbit and on our home planet.

CHAPTER V
LUNAR EXPLORATION AND HABITATION

LUNAR OUTPOSTS: ESTABLISHING MOON BASES

Lunar outposts represent the first step towards establishing a sustainable human presence on the Moon. These bases will serve as hubs for scientific research, resource utilization, and as staging points for further space exploration. The development of lunar outposts involves significant challenges, including the harsh lunar environment, the need for sustainable life support systems, and the logistics of transporting materials and personnel.

Key Players and initiatives:

1. **NASA:** Leading the Artemis program, which aims to establish a sustainable presence on the Moon by the end of the decade. The Artemis Base Camp will be located near the lunar South Pole, leveraging the presence of water ice for life support and fuel production.
2. **Lunar Outpost:** A company dedicated to providing innovative technology and services to enable a sustainable human presence in space. They are involved in developing advanced spacecraft and robotics solutions for lunar missions.
3. **SpaceX:** Developing the Starship for lunar missions, including the Human Landing System (HLS) for NASA's Artemis program.
4. **Blue Origin:** Working on the Blue Moon lander and collaborating with other companies to develop lunar infrastructure.
5. **Intuitive Machines and Venturi Astrolab:** Selected by NASA to advance capabilities for lunar terrain vehicles (LTVs) that will support Artemis missions.

ARTEMIS BASE CAMP:
NASA'S VISION FOR LUNAR HABITATION

NASA's Artemis Base Camp is a cornerstone of the agency's vision for sustained lunar exploration. The base camp will be located near the Moon's South Pole, where sunlight is abundant and water ice is accessible. The Artemis Base Camp will include several key components:

1. **Foundation Surface Habitat:** A habitat designed for short-term crew stays, providing living and working space for astronauts.
2. Habitable Mobility Platform: A mobile habitat that allows astronauts to conduct long-duration missions away from the base camp.
3. **Lunar Terrain Vehicle (LTV):** An unpressurized rover for transportation and exploration on the lunar surface.
4. **Power and Communication Infrastructure:** Solar power systems and communication networks to support the base camp's operations.

MINING THE MOON: LUNAR RESOURCES AND UTILIZATION

The Moon's surface is rich in resources that can support human habitation and further space exploration. The development of technologies to extract and use these resources, known as in-situ resource utilization (ISRU), is critical for establishing a sustainable presence on the Moon and beyond. Key resources include water ice, regolith, and helium-3, each offering unique benefits and challenges.

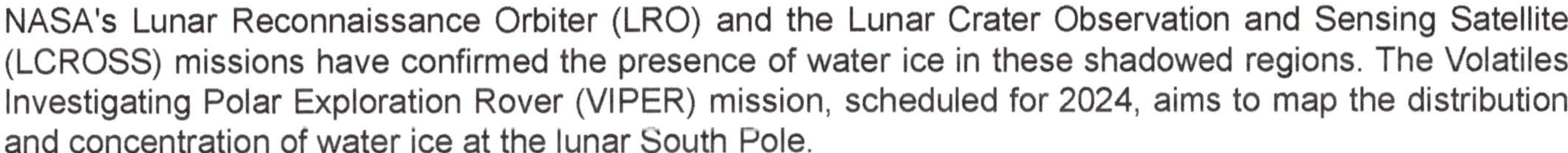

Water Ice

Water ice is one of the most valuable resources on the Moon, primarily found in permanently shadowed craters at the lunar poles. This ice can be used for:

1. **Life Support:** Water is essential for drinking, hygiene, and food preparation. It can also be split into hydrogen and oxygen through electrolysis, providing breathable oxygen and hydrogen for rocket fuel.
2. **Radiation Shielding:** Water can be used as a radiation shield to protect habitats and equipment from harmful cosmic rays and solar radiation.
3. **Agriculture:** Water is necessary for growing food in hydroponic and aeroponic systems, supporting long-term human habitation.

NASA's Lunar Reconnaissance Orbiter (LRO) and the Lunar Crater Observation and Sensing Satellite (LCROSS) missions have confirmed the presence of water ice in these shadowed regions. The Volatiles Investigating Polar Exploration Rover (VIPER) mission, scheduled for 2024, aims to map the distribution and concentration of water ice at the lunar South Pole.

Regolith

Lunar regolith, or lunar soil, is a versatile resource that can be processed to extract oxygen, metals, and other valuable materials. Key applications include:

1. Oxygen Production: Regolith contains oxides that can be processed to extract oxygen, which is crucial for life support and rocket propellant. Techniques such as molten regolith electrolysis and carbothermal reduction are being developed to achieve this.

2. Construction Materials: Regolith can be used to create building materials for habitats, roads, and other infrastructure. 3D printing technologies, such as those developed by ICON and AI SpaceFactory, can use regolith to construct durable structures on-site.

3. Radiation Shielding: Regolith can be used to cover habitats and other structures, providing protection from radiation and micrometeorite impacts.

The European Space Agency (ESA) and NASA are actively researching methods to process regolith for these purposes. ESA's Regolith Advanced Surface Systems Operations Robot (RASSOR) and NASA's Carbothermal Reduction Demonstration (CaRD) are examples of ongoing projects aimed at developing efficient regolith processing technologies.

Helium-3

Helium-3 is a rare isotope found in the lunar regolith, deposited by the solar wind over billions of years. It has significant potential as a fuel for future nuclear fusion reactors due to its unique properties:

1. **Clean Energy:** Helium-3 fusion produces minimal radioactive waste compared to traditional nuclear fusion, making it a cleaner energy source.

2. **High Energy Yield:** Fusion of helium-3 with deuterium produces a high energy yield, which could provide a substantial power source for lunar bases and future space missions.

China's Chang'e missions and NASA's Artemis program are investigating the feasibility of extracting helium-3 from the lunar regolith. Companies like Interlune and Helion Energy are exploring the commercial potential of helium-3 mining and its use in fusion reactors.

Challenges and Considerations

While the potential benefits of lunar resource utilization are significant, several challenges must be addressed:

1. Technological Development: Advanced technologies are required to extract, process, and utilize lunar resources efficiently. This includes developing robust mining equipment, refining extraction techniques, and creating sustainable processing systems.

2. Economic Viability: The high cost of lunar missions necessitates a clear economic justification for resource extraction. This includes demonstrating the value of lunar resources compared to Earth-based alternatives.

3. Environmental Impact: Mining activities must be conducted responsibly to minimize the environmental impact on the lunar surface. This includes managing lunar dust, preserving scientific sites, and ensuring sustainable practices.

4. Regulatory Framework: International agreements and regulations are needed to govern the extraction and use of lunar resources. This includes defining property rights, ensuring fair access, and preventing conflicts.

The key technologies needed to establish sustainable lunar outposts include:

1. **Life Support Systems**
2. **Power Generation and Storage**
3. **In-Situ Resource Utilization (ISRU)**
4. **Habitat Construction**
5. **Food Production**
6. **Communication Systems**
7. **Transportation and Mobility**
8. **Health and Medical Technologies**
9. **Robotics and Automation**
10. **Dust Mitigation**

SPACEPOSIUM

These technologies need to be highly reliable, efficient, and designed to operate in the harsh lunar environment with minimal resupply from Earth. Developing integrated systems that can utilize lunar resources and operate with a high degree of autonomy will be crucial for establishing sustainable long-term outposts on the Moon.

Mining the Moon's resources is a critical step towards establishing a sustainable human presence on the lunar surface and beyond. By developing technologies to extract and utilize water ice, regolith, and helium-3, we can reduce the reliance on Earth-based supplies, lower mission costs, and enable long-term exploration and habitation. The collaborative efforts of space agencies, private companies, and international partners will be essential in overcoming the challenges and realizing the full potential of lunar resource utilization. As we continue to explore and develop the Moon, the knowledge and technologies gained will pave the way for future missions to Mars and other destinations in our solar system.

Conclusion

Lunar exploration and habitation represent a new frontier in human space exploration. Establishing a sustainable presence on the Moon will require collaboration between space agencies, private companies, and international partners. The development of lunar outposts, resource utilization, and innovative construction methods will pave the way for long-term human habitation on the Moon and serve as a stepping stone for future missions to Mars and beyond. As we embark on this ambitious journey, the lessons learned and technologies developed will not only advance our understanding of the Moon but also provide valuable insights and innovations for life on Earth.

Top companies working in lunar mining operations, resource processing, and construction:

1. Lunar Resources, Inc. - Focused on space manufacturing, off-earth resource extraction, and in-situ resource utilization.
2. Moon Express - Developing robotic spacecraft for lunar exploration and resource utilization.
3. Interlune - Aims to harvest Helium-3 from the Moon for various applications.
4. Argo Space Corp - Specializes in harvesting water from lunar regolith for propellant production.
5. CisLunar Industries - Developing metal processing capabilities for space-based operations.
6. Off Planet Research - Creates lunar regolith simulants for testing space technologies.
7. iSpace - Developing lunar landers and rovers for exploration and resource utilization.

8. Astrobotic Technology - Building lunar landers and rovers for NASA's Commercial Lunar Payload Services program.
9. Blue Origin - Developing the Blue Moon lander for lunar cargo and crewed missions.
10. ICON - Focused on 3D printing technologies for lunar construction.
11. Masten Space Systems - Developing lunar landers and ISRU technologies.
12. Honeybee Robotics - Creating drilling and sampling systems for lunar resource extraction.
13. Redwire Space - Developing various space technologies, including those for lunar operations.
14. TransAstra Corporation - Working on optical mining technology for resource extraction.
15. Karman+ - Developing technologies for mining water from near-Earth asteroids (potentially applicable to lunar mining).

These companies are involved in various aspects of lunar operations, from resource identification and extraction to processing and construction technologies needed for establishing a sustainable presence on the Moon.

BUILDING ON THE MOON: HABITATS AND CONSTRUCTION

Constructing habitats on the Moon presents unique challenges due to the harsh lunar environment, including extreme temperature variations, radiation, micrometeorite impacts, and the abrasive nature of lunar dust. As we prepare for long-term lunar exploration and habitation, innovative approaches and technologies are being developed to address these challenges.

Key Construction Methods and Technologies
3D Printing with Lunar Regolith:

3D printing with lunar regolith is emerging as a groundbreaking technology for lunar construction. Companies like ICON and AI SpaceFactory are at the forefront, developing innovative 3D printing technologies that utilize lunar regolith as a raw material. This approach offers a significant advantage by enabling on-site construction, dramatically reducing the need to transport building materials from Earth. NASA is also investing in this technology through its MMPACT (Moon-to-Mars Planetary Autonomous Construction Technology) project, which aims to use 3D printing for building lunar infrastructure. This method could revolutionize how we approach construction in extraterrestrial environments, making long-term lunar habitation more feasible and sustainable.

Inflatable Habitats:

Due to their lightweight and compact nature, inflatable habitats represent a promising solution for lunar construction. These structures can be easily transported to the Moon and expanded on-site, providing a quick and efficient method for establishing living quarters. To enhance their protective capabilities, these inflatable habitats can be covered with lunar regolith, offering additional shielding against radiation and micrometeorites. Companies like Bigelow Aerospace have been at the forefront of developing inflatable habitat concepts for space applications, demonstrating the potential of this technology for future lunar missions and long-term habitation.

Robotic Construction:

Robotic construction is set to play a crucial role in lunar development, offering a safer and more efficient alternative to human labor in the hazardous lunar environment. Autonomous robots can perform a wide range of construction tasks, reducing the risks associated with human EVAs (Extra-Vehicular Activities). Companies like GITAI are leading the way in developing multi-purpose robots specifically designed for lunar construction and maintenance. Similarly, OffWorld is working on advanced robotic systems capable of extracting and processing lunar resources, as well as constructing structures on the Moon. These robotic systems will be essential for establishing and maintaining a sustainable human presence on the lunar surface.

INNOVATIVE HABITAT DESIGNS FOR LUNAR EXPLORATION AND HABITATION

NASA's Artemis Base Camp:
NASA's Artemis Base Camp is a planned permanent outpost at the lunar South Pole, designed to support long-term human presence on the Moon. The location near the South Pole is strategic, offering access to areas of near-constant sunlight for power generation and potentially water ice in permanently shadowed craters.

ESA's Moon Village Concept:
The European Space Agency's Moon Village is not a single project but a vision for international cooperation in lunar exploration and utilization. This is to be a combination of human and robotic presence: The concept envisions both crewed missions and autonomous systems working together.

ICON's Project Olympus:

ICON, known for its 3D-printed housing on Earth, is developing advanced construction technologies for lunar applications. Project Olympus includes large-scale 3D printing system designed to use lunar regolith as a building material. Multi-purpose construction capabilities: Aimed at creating various structures including landing pads, roads, and habitats. ICON has received significant NASA funding to further develop this technology, with the goal of supporting NASA's Artemis program and future commercial lunar activities

Foster + Partners' Lunar Habitation:
The renowned architecture firm has developed a concept for 3D-printed lunar habitats in collaboration with the European Space Agency. Key aspects include an inflatable core, consisting of a transportable, inflatable structure forms the base of the habitat. Aregolith shell: Robots would 3D print a protective shell using lunar soil over the inflatable core. And a cellular structure: The printed shell would have a hollow, closed-cell structure inspired by biological forms for strength and efficiency.

These innovative habitat designs represent different approaches to the challenge of lunar habitation, from government-led initiatives to commercial ventures and architectural concepts. They all share a focus on utilizing lunar resources, protecting inhabitants from the harsh environment, and creating sustainable, expandable structures for long-term lunar presence.

CONCLUSION: PAVING THE WAY FOR HUMANITY'S LUNAR FUTURE

As we stand on the brink of a new era in lunar exploration and habitation, it's clear that the

challenges ahead are as numerous as they are complex. From establishing sustainable lunar outposts to mining lunar resources and constructing habitats in the harsh lunar environment, each aspect of this endeavor pushes the boundaries of human ingenuity and technological innovation.

The collaborative efforts of space agencies, private companies, and international partners are driving rapid advancements in key areas such as in-situ resource utilization, 3D printing with lunar regolith, and the development of inflatable and robotic construction techniques. These innovations are not just solving immediate challenges but are laying the groundwork for long-term human presence on the Moon.

The focus on utilizing lunar resources, from water ice at the poles to metals and oxygen in the regolith, underscores the potential for the Moon to serve as a proving ground for technologies and techniques that will be crucial for future Mars missions and beyond. At the same time, the emphasis on responsible and ethical lunar exploration reflects a growing awareness of the need to preserve the Moon's scientific and cultural value.

As we move forward, the challenges of lunar exploration and habitation will continue to drive innovation in fields ranging from materials science and robotics to life support systems and space medicine. The lessons learned and technologies developed will not only enable our sustained presence on the Moon but will also have far-reaching implications for life on Earth, potentially leading to breakthroughs in areas such as sustainable living, resource management, and remote operations.

The journey to establish a human presence on the Moon is more than just a technical challenge; it represents a new chapter in human history. As we take these next steps, we do so with a sense of responsibility to explore ethically, to cooperate internationally, and to use our ingenuity to overcome the formidable challenges that lie ahead. The Moon, our celestial neighbor, stands not just as a destination, but as a gateway to the broader solar system and a testament to human determination and innovation.

SPACEPOSIUM

CHAPTER VI
MARS EXPLORATION AND COLONIZATION

In the vast expanse of our cosmic neighborhood, a rusty jewel beckons humanity with an irresistible siren song. Mars, the Red Planet, has ignited our collective imagination for millennia, its ruddy glow a constant reminder of unexplored frontiers and untold mysteries. Now, as we stand at the precipice of a new era in human history, Mars looms larger than ever before – not just as a distant celestial body, but as the next great leap for our species.

This chapter embarks on an epic journey through the realm of Mars exploration and colonization, where science fiction meets cutting-edge reality. We'll traverse the treacherous terrain of Martian geology, unravel the secrets hidden in its dusty atmosphere, and gaze into the abyss of time to uncover the planet's tumultuous past. From the first robotic emissaries that touched its alien soil to the audacious plans for establishing a permanent human foothold, we'll chronicle humanity's relentless pursuit of the Red Planet.

As we contemplate the monumental task of colonizing Mars, we'll grapple with profound questions about our species' destiny and our place in the cosmos. The Red Planet stands as both a mirror to our past and a gateway to our future – a testament to human ingenuity and an ultimate test of our resolve

Strap in for an odyssey that spans the gulf between worlds, as we set our sights on the most audacious endeavor in human history: the conquest of Mars.

From understanding the Martian environment to planning long-term colonies, from the grand vision of terraforming to the practical challenges of living in an alien world, we will explore the cutting-edge science, technology, and human ingenuity driving our journey to Mars. As we embark on this exploration, we'll examine the roles of key players such as NASA, ESA, SpaceX, and other international space agencies and private companies in shaping the future of Mars exploration and colonization.

The quest to explore and potentially colonize Mars represents one of the most ambitious endeavors in human history. It pushes the boundaries of our scientific knowledge, technological capabilities, and our understanding of what it means to be human in the cosmos. As we progress through this chapter, we'll uncover the challenges, opportunities, and potential benefits that Mars exploration offers not just for space exploration, but for life on Earth as well.

Let's journey through the current state of Mars exploration, peer into the future of Martian colonies, grapple with the monumental task of terraforming an entire planet, and envision what daily life might be like for the first Martians. This is the story of humanity's next great leap – the exploration and colonization of Mars.

THE RED PLANET: UNDERSTANDING MARS

Mars, often called the Red Planet due to its reddish appearance, is the fourth planet from the Sun and Earth's closest planetary neighbor. Key points to understand about Mars include:

- Physical characteristics: Mars is about half the size of Earth, with a thin atmosphere composed mainly of carbon dioxide. It has a cold, desert-like surface with polar ice caps.
- Geological features: Mars has the largest volcano in the solar system (Olympus Mons) and a canyon system (Valles Marineris) that dwarfs Earth's Grand Canyon.
- Water on Mars: Evidence suggests Mars once had liquid water on its surface and still has water ice at its poles and in subsurface deposits.
- Martian day and year: A Martian day (sol) is slightly longer than Earth's at 24 hours and 37 minutes. A Martian year lasts about 687 Earth days.
- Moons: Mars has two small, irregularly shaped moons - Phobos and Deimos.

Understanding these characteristics is crucial for planning human missions and potential colonization efforts.

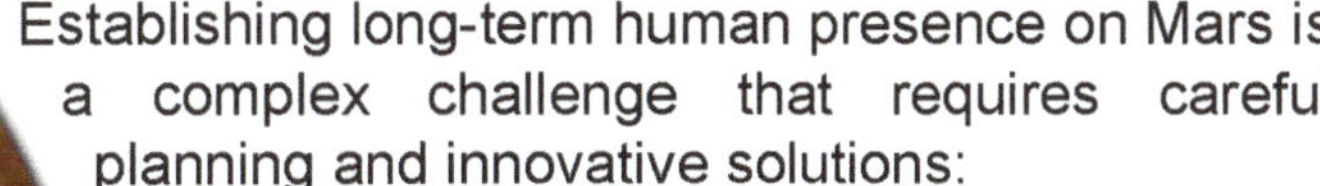

Establishing long-term human presence on Mars is a complex challenge that requires careful planning and innovative solutions:

- Habitat design: Concepts include inflatable structures, 3D-printed habitats using Martian regolith, and the use of lava tubes for natural radiation shielding.
- Life support systems: Developing closed-loop systems for air, water, and waste recycling is crucial for long-term sustainability.
- Energy production: Solar power, nuclear power, and potentially geothermal energy are being considered for Mars colonies.
- Psychological considerations: Addressing isolation, confinement, and extreme environments through crew selection and habitat design.
- Transportation systems: Developing reliable systems for Earth-Mars transport and surface mobility on Mars.
- In-situ resource utilization (ISRU): Plans to use Martian resources for fuel, water, and building materials to reduce reliance on Earth supplies.

Companies like SpaceX, Blue Origin, and NASA are actively developing technologies and plans for Mars colonization.

Terraforming Mars: Possibilities and Challenges

Terraforming Mars involves transforming the planet to make it more Earth-like and habitable for humans. While currently beyond our technological capabilities, key considerations include:

- Increasing atmospheric pressure: Mars' atmosphere is too thin to support liquid water or human life without pressure suits.
- Warming the planet: Proposals include releasing greenhouse gases from polar ice caps or introducing engineered microorganisms.
- Creating a magnetic field: Mars lacks a strong magnetic field, leaving it vulnerable to solar radiation.
- Ethical considerations: Debates about the right to alter another planet's environment and the potential impact on existing microbial life.
- Timeframe: Even with advanced technology, terraforming would likely take centuries to achieve meaningful results.

Current research suggests that terraforming Mars is not feasible with present-day technology, but it remains a topic of scientific interest and speculation.

Living on Mars: Resources, Agriculture, and Adaptation

Sustaining human life on Mars will require innovative approaches to resource management, food production, and human adaptation:

- Water extraction: Techniques for extracting water from subsurface ice deposits and the atmosphere are being developed.
- Agriculture: Experiments with growing food in simulated Martian soil and controlled environment agriculture (CEA) are ongoing. NASA's VEGGIE project on the ISS is testing food production in space.
- Radiation protection: Developing habitats and spacesuits to shield against high levels of cosmic radiation.
- Health considerations: Addressing the effects of reduced gravity on the human body, including bone and muscle loss.
- Psychological well-being: Creating environments that support mental health through design, activities, and connection with Earth.
- Resource recycling: Implementing advanced recycling systems for air, water, and waste to minimize reliance on resupply missions.

Companies like ICON and AI SpaceFactory are developing 3D printing technologies for Martian habitats, while others like Interstellar Lab focus on closed-loop life support systems.

This expanded overview provides a comprehensive look at the challenges and opportunities involved in Mars exploration and colonization, highlighting the interdisciplinary nature of this ambitious endeavor.

Private companies, particularly SpaceX, play a crucial role in advancing Mars colonization through several key contributions:

Technological Innovation and Cost Reduction

SpaceX has revolutionized space travel with its development of reusable rocket technology, significantly reducing the cost of space access. The Falcon 9 and Falcon Heavy rockets, along with the Starship spacecraft, are designed to be reusable, which lowers the cost of launching payloads and humans into space. This cost efficiency is essential for the feasibility of long-term Mars missions.

Ambitious Vision and Goals

Elon Musk, the founder of SpaceX, has articulated a clear and ambitious vision for making humanity a multi-planetary species. SpaceX aims to establish a self-sustaining human colony on Mars, with plans to transport large numbers of people and cargo to the Red Planet using the Starship spacecraft. This vision drives the company's efforts and inspires public and private sector collaboration.

Collaboration with Government Agencies

SpaceX collaborates closely with NASA and other governmental space agencies. These partnerships leverage the strengths of both public and private sectors, combining NASA's expertise and long-term research capabilities with SpaceX's innovative approaches and agility. Such collaborations are crucial for the success of complex missions like Mars colonization.

Infrastructure Development

SpaceX is not only focused on transportation but also on developing the necessary infrastructure for Mars missions. This includes landing pads, refueling stations, and life-support systems that will be essential for sustaining human life on Mars. The company plans to use in-situ resource utilization (ISRU) techniques to produce fuel on Mars from local resources, which is vital for return missions and long-term sustainability.

Pioneering Milestones

SpaceX has achieved several historic milestones that demonstrate its capability and reliability. These include the first privately funded spacecraft to reach orbit, the first private company to send a spacecraft to the International Space Station, and the first private company to launch and land reusable rockets. These achievements build confidence in SpaceX's ability to undertake the monumental task of Mars colonization.

Public Engagement and Inspiration

SpaceX's high-profile missions and public engagement strategies have generated significant public interest and support for space exploration. By live-streaming launches and sharing progress updates, SpaceX has inspired a new generation of scientists, engineers, and space enthusiasts, fostering a culture of innovation and exploration.

Private companies like SpaceX are at the forefront of the push towards Mars colonization, bringing innovation, reducing costs, and fostering collaboration with governmental agencies. Their ambitious goals, technological advancements, and public engagement efforts are critical in making the dream of a human presence on Mars a reality. As we look to the future, the role of private companies will continue to be pivotal in overcoming the challenges and seizing the opportunities that Mars exploration presents.

How can virtual reality help Mars settlers cope with isolation?

There are several critical ways that virtual reality (VR) could potentially help Mars settlers cope with isolation:

• Providing immersive social experiences: VR can allow settlers to feel connected to others and have social interactions, even when physically isolated. Social VR platforms enable people to meet and interact in virtual spaces, potentially reducing feelings of loneliness.

• Simulating Earth environments: VR can transport settlers to familiar Earth environments, allowing them to virtually experience nature, visit their hometowns, or attend events, helping combat homesickness and providing a mental escape.

• Eliciting positive emotions: Studies have shown VR experiences of awe-inspiring scenery or flying around Earth can produce positive emotions, feelings of connection, and perspective changes that could be beneficial for mental health.

• Offering novel experiences: VR can provide settlers with new and stimulating experiences to combat boredom and monotony, keeping their minds engaged.

• Facilitating communication with Earth: VR could potentially enhance communication with family and friends on Earth by creating more immersive shared experiences despite the physical distance.

• Supporting mental health interventions: VR has shown promise as a tool for delivering therapeutic interventions and improving social well-being, which could be adapted for settlers' needs.

• Providing entertainment and recreation: VR games and experiences can offer much-needed entertainment and recreational activities in the confined Mars environment.

• Assisting with mission tasks: AR/VR tools can help with training, maintenance tasks, and robot control, potentially reducing stress associated with complex mission activities.

While VR shows promise for helping with isolation, its long-term effects are still being studied. It should be used as part of a comprehensive strategy to support settlers' mental health and well-being during a Mars mission.

Several VR companies stand out as potentially capable of handling this type of project. Here are some of the best VR companies that could potentially address this challenge, along with their names and URLs:

1. AppliedVR (https://appliedvr.io/)
AppliedVR specializes in therapeutic VR applications, particularly for pain management and mental health. Their expertise in creating immersive experiences for healthcare could be valuable for developing solutions to help Mars settlers manage stress and isolation.

2. Osso VR (https://www.ossovr.com/)
While primarily focused on surgical training, Osso VR's experience in creating highly detailed and interactive VR environments could be adapted to create realistic Earth-like experiences for Mars settlers.

3. Unity Technologies (https://unity.com/)
Unity is a leading platform for creating and operating interactive, real-time 3D content. Their powerful engine could be used to develop a wide range of immersive experiences tailored to the needs of Mars settlers.
4. Meta (formerly Oculus VR) (https://www.meta.com/)
As a pioneer in VR technology, Meta (formerly Facebook's Oculus) has extensive experience in creating social VR platforms and high-quality VR hardware. Their expertise could be crucial in developing systems for Mars settlers to connect with Earth and each other.
5. HTC VIVE (https://www.vive.com/)
HTC VIVE offers advanced VR and AR solutions that could be adapted for use in a Mars settlement. Their experience in creating immersive environments and developing VR hardware could be valuable for this challenge.
6. Magic Leap (https://www.magicleap.com/)
Magic Leap specializes in mixed reality technology, which could be particularly useful for creating augmented experiences within the Mars habitat, potentially helping settlers feel less confined.
7. Varjo (https://varjo.com/)
Varjo produces high-end VR and XR headsets with human-eye resolution. Their advanced technology could be crucial for creating ultra-realistic virtual experiences for Mars settlers.

These companies have demonstrated expertise in creating immersive VR experiences and have the technological capabilities to potentially address the unique challenges of helping Mars settlers cope with isolation. However, it's important to note that such a project would likely require collaboration between multiple companies and space agencies to fully address the complex needs of a Mars mission.

Conclusion Mars Exploration and Colonization:

The exploration and potential colonization of Mars represent one of humanity's greatest challenges and opportunities. As we've seen throughout this chapter, the journey to establish a human presence on the Red Planet is fraught with immense technical, logistical, and physiological hurdles. Yet the potential rewards—scientific discovery, technological innovation, and the expansion of human civilization beyond Earth—are equally immense.

From the early robotic missions that have mapped Mars' surface and analyzed its composition to the ambitious plans for human missions in the coming decades, we are witnessing the gradual unfolding of a new era in space exploration. The collaborative efforts of space agencies, private companies, and international partners are driving progress at an unprecedented pace.

However, as we look towards a future where humans may live and work on Mars, we must also grapple with profound ethical, legal, and philosophical questions. How will we ensure the responsible stewardship of a new world? What impact might our presence have on potential

Martian life? How will Martian societies be governed? The journey to Mars is not just about reaching another planet – it's about expanding the horizons of human knowledge, capability, and imagination. Every step we take towards Mars pushes the boundaries of science and engineering, with potential benefits that extend far beyond space exploration. The technologies and solutions developed for Mars missions could help address pressing challenges on Earth, from sustainable resource utilization to climate change mitigation.

As we conclude this chapter, it's clear that the road to Mars is long and challenging, but the destination holds the promise of transforming humanity into a multi-planetary species. The exploration and colonization of Mars may well be the defining adventure of the 21st century and beyond, opening up new frontiers for human achievement and understanding. As we continue to push forward, we carry with us the hopes, dreams, and ingenuity of our entire species, reaching for the stars while keeping our feet firmly planted in the red Martian soil.

CHAPTER VII
EARTH OBSERVATION

As humanity extends its reach into space, our ability to observe Earth and utilize extraterrestrial resources is rapidly evolving. This chapter explores the cutting-edge technologies and applications that are reshaping our understanding of our planet and opening new frontiers for resource acquisition beyond Earth. From advanced Earth observation satellites to the promising field of asteroid mining and the potential of directed energy in space, we delve into the innovations that are not only providing crucial data about our changing world but also paving the way for sustainable space exploration and resource utilization.

CHARTING THE SKIES: EARTH OBSERVATION AND ITS IMPACT

Earth observation (EO) technology has undergone remarkable advancements in recent years, revolutionizing our ability to monitor and understand our planet. This section explores the current state of Earth observation, its applications, and the profound impact it's having across various sectors.

Technological Advancements:

The Earth observation field has seen significant technological progress, particularly in satellite technology and data processing capabilities:

1. High-resolution imaging: Modern EO satellites can capture images with resolutions as fine as 30 cm, allowing for unprecedented detail in Earth monitoring.
2. Increased revisit rates: Constellations of small satellites now provide near-real-time monitoring of the Earth's surface, with some areas imaged multiple times per day.
3. Multi-spectral and hyperspectral sensors: These advanced sensors capture data across a wide range of the electromagnetic spectrum, enabling detailed analysis of Earth's features and processes.
4. Synthetic Aperture Radar (SAR): This technology allows for Earth observation regardless of weather conditions or time of day, greatly expanding monitoring capabilities.
5. AI and machine learning: These technologies are increasingly used to process and analyze the vast amounts of data generated by EO satellites, extracting valuable insights more efficiently than ever before.

Applications and Impact:

1. Climate monitoring: EO data is crucial for tracking global temperature changes, sea level rise, ice melt, and other climate change indicators, informing policy decisions and climate action.
2. Disaster management: Satellite imagery provides critical information for disaster response and recovery efforts, from mapping flood extents to assessing wildfire damage.
3. Agriculture: EO data helps optimize crop yields, monitor soil health, and predict harvests on a global scale, contributing to food security efforts.
4. Urban planning: High-resolution imagery and 3D modeling support sustainable urban development, infrastructure planning, and monitoring of urban growth.
5. Environmental protection: Satellites track deforestation, monitor air and water quality, and help enforce environmental regulations.
6. Maritime surveillance: EO technology aids in monitoring shipping routes, detecting illegal fishing, and tracking oil spills.
7. Defense and security: Satellite imagery plays a crucial role in national security, border monitoring, and conflict assessment.

What is Hyperspectral Satellite Technology?

Hyperspectral satellite technology is an advanced form of Earth observation that captures images across a wide range of the electromagnetic spectrum. Unlike traditional imaging systems that capture data in a few broad spectral bands, hyperspectral imaging collects data in hundreds of narrow, contiguous spectral bands. This allows for the creation of detailed spectral signatures for each pixel in an image, enabling the identification and analysis of materials and objects with high precision. The technology is used in various applications, including environmental monitoring, mineral exploration, agriculture, and defense, providing critical insights that are not possible with conventional imaging methods.

Several companies are leading the development of hyperspectral satellite technology. Pixxel (https://pixxel.space/) is building a constellation of hyperspectral Earth-imaging satellites, partnering with organizations like NASA, JPL, and Lockheed Martin. Planet Labs (https://www.planet.com/) is developing hyperspectral capabilities with their Tanager satellites.

Orbital Sidekick (https://orbitalsidekick.com/) focuses on remote monitoring for the energy, mining, and defense industries with their hyperspectral sensors. Other notable companies include Gamaya (https://gamaya.com/) for precision agriculture and Imec (https://www.imechyperspectral.com/en/applications/hyperspectral-remote-sensing) for compact hyperspectral sensors used in various applications. These companies are at the forefront of advancing hyperspectral imaging technology, making it more accessible and applicable across different industries.

The Growing Commercial Market:

The Earth observation market has seen significant growth and commercialization in recent years:

1. Market size: The global EO market was valued at $3.3 billion in 2023 and is projected to reach $7.9 billion by 2030, growing at a CAGR of 13.2% from 2024 to 2030.
2. Key players: Major companies in the commercial EO sector include:Maxar Technologies
3. Democratization of data: Many companies now offer easy access to EO data through cloud-based platforms, making it more accessible to a wider range of users and applications.
4. Value-added services: There's a growing market for companies that provide analysis and insights based on EO data, rather than just raw imagery.

Challenges and Future Outlook:

Despite its rapid growth, the EO sector faces challenges:
1. Data overload: The sheer volume of data generated by EO satellites can be overwhelming, necessitating advanced data management and analysis techniques.
2. Privacy concerns: High-resolution imaging raises questions about privacy and security.
3. Space debris: The increasing number of satellites in orbit contributes to the growing problem of space debris.
4. Standardization: There's a need for better standardization of data formats and analysis methods across the industry.

Looking ahead, the Earth observation field is poised for continued growth and innovation. Emerging technologies like quantum sensors, artificial intelligence, and edge computing are expected to enhance further our ability to monitor and understand our planet, providing crucial data for addressing global challenges like climate change, food security, and sustainable development.

Advancements in Earth observation satellite technology

Advancements in Earth observation satellite technology have revolutionized our ability to monitor and understand our planet. Here are some key developments:
1. Higher Resolution Imaging: Modern satellites can capture images with resolutions as fine as 30 cm, allowing for unprecedented detail in Earth monitoring. Companies like Maxar Technologies (https://www.maxar.com/) are at the forefront of high-resolution imaging.
2. Increased Revisit Rates: Constellations of small satellites now provide near-real-time monitoring of the Earth's surface, with some areas imaged multiple times per day. Planet Labs (https://www.planet.com/) is a leader in this area with their large constellation of small satellites.
3. Synthetic Aperture Radar (SAR): This technology allows for Earth observation regardless of weather conditions or time of day. ICEYE (https://www.iceye.com/) and Capella Space (https://www.capellaspace.com/) are pioneering commercial SAR satellite constellations.
4. Hyperspectral Imaging: These sensors capture data across a wide range of the electromagnetic spectrum, providing detailed information about the composition of objects on Earth. Companies like Satellogic (https://satellogic.com/) are developing hyperspectral imaging capabilities.
5. AI and Machine Learning Integration: Advanced algorithms are being used to process and analyze the vast amounts of data generated by EO satellites, extracting valuable insights more efficiently. BlackSky (https://www.blacksky.com/) is integrating AI into their satellite imagery analysis.
6. Miniaturization: The development of smaller, more efficient satellites has reduced launch costs and increased the accessibility of space-based Earth observation. Spire Global (https://spire.com/) specializes in nanosatellites for Earth observation.
7. Improved Temporal Resolution: More frequent revisits to the same location allow for better monitoring of rapidly changing phenomena. Airbus Defence and Space (https://www.airbus.com/en/products-services/space) offers high temporal resolution through their constellation.

8. Multi-sensor Integration: Combining data from multiple types of sensors (optical, radar, infrared, etc.) provides a more comprehensive view of Earth's systems. The European Space Agency's Copernicus program (https://www.copernicus.eu/en) exemplifies this approach.

These advancements are enabling more accurate, timely, and comprehensive Earth observation, supporting applications in climate monitoring, disaster management, urban planning, agriculture, and many other fields.

TOP COMPANIES IN THE FIELD

Here are some of the top companies leading advancements in Earth observation satellite technology include:

1. Maxar Technologies (https://www.maxar.com) - Provides high-resolution satellite imagery and geospatial solutions.
2. Planet Labs (https://www.planet.com) - Operates a large constellation of small satellites for frequent Earth imaging.
3. ICEYE (https://www.iceye.com) - Specializes in SAR (synthetic aperture radar) satellite technology for all-weather Earth observation.
4. Airbus Defence and Space (https://www.airbus.com/en/products-services/space) - Develops advanced Earth observation satellites and services.
5. BlackSky (https://www.blacksky.com) - Offers real-time geospatial intelligence and monitoring.
6. Capella Space (https://www.capellaspace.com) - Provides on-demand SAR satellite imagery.
7. Satellogic (https://satellogic.com) - Develops high-resolution multispectral imaging satellites.
8. Spire Global (https://spire.com) - Operates nanosatellites for Earth observation and tracking.
9. OroraTech (https://www.ororatech.com) - Specializes in wildfire detection and monitoring using satellite data.
10. LiveEO (https://www.live-eo.com) - Uses AI and satellite data for infrastructure monitoring.

These companies are at the forefront of developing advanced satellite technologies, improving image resolution, increasing revisit rates, expanding spectral capabilities, and applying AI/ML to extract insights from satellite data. With their Earth observation solutions, they serve various industries, including agriculture, forestry, energy, defense, and disaster response.

ASTEROID MINING: HARVESTING CELESTIAL TREASURE

Asteroid mining represents one of the most ambitious and potentially transformative endeavors in the realm of space exploration and resource utilization. This section explores the immense potential of asteroid resources, the current state of asteroid mining technology, and the companies leading the charge in this emerging field.

The Potential of Asteroid Resources:

1. Metals: Asteroids are rich in valuable metals, including:
- Platinum group metals (platinum, palladium, rhodium)
- Gold and silver
- Iron, nickel, and cobalt

Some asteroids are believed to contain higher concentrations of these metals than Earth's richest mines. For example, a single 500-meter-wide platinum-rich asteroid could contain nearly 175 times the global annual output or 1.5 times the known world reserves of platinum group metals.

2. Water: Many asteroids contain significant amounts of water ice, which is crucial for:
- Life support systems in space
- Producing rocket fuel (through electrolysis into hydrogen and oxygen)
- Radiation shielding for spacecraft

3. Rare Earth Elements: Asteroids may contain concentrations of rare earth elements essential for modern electronics and green technologies.

Rare earth elements (REEs) found in asteroids could indeed be a valuable resource for modern technology and green energy applications. Here's an elaboration on this point:

Types of REEs in asteroids: Asteroids, particularly certain types like carbonaceous chondrites, may contain higher concentrations of REEs than Earth's crust. These could include elements such as neodymium, dysprosium, terbium, europium, and yttrium.

Importance for technology: REEs are crucial for many high-tech applications due to their unique magnetic, phosphorescent, and catalytic properties. They're used in:
- Smartphones and computer components
- Electric and hybrid vehicle motors
- Wind turbine generators
- LED lights and displays
- Laser technologies
- Catalytic converters

Green energy applications: REEs are particularly important for clean energy technologies:
   - Neodymium and dysprosium for powerful permanent magnets in wind turbines and electric vehicles
   - Europium, terbium, and yttrium for energy-efficient lighting
   - Lanthanum and cerium for catalysts in fuel cells
Potential advantages of asteroid mining:
   - Reduced environmental impact compared to terrestrial mining
   - Potentially higher ore grades than found on Earth
   - New supply chains independent of geopolitical tensions

Challenges:
   - High initial costs for space missions
   - Technological hurdles in extraction and processing in space
   - Need for advances in in-situ resource utilization (ISRU) technologies

Current research: While no asteroid mining operations are currently active, several companies and space agencies are researching the potential, including developing technologies for prospecting and extracting resources from asteroids.

Long-term potential: As terrestrial supplies of some REEs become strained, asteroid mining could become an important supplementary source, potentially revolutionizing the supply chain for these critical elements.

It's important to note that while the potential is there, asteroid mining for REEs remains speculative at this point. Significant technological and economic challenges need to be overcome before it becomes a reality.

SPACEPOSIUM

Current Challenges:
1. Technological: Developing space-rated mining and processing equipment capable of operating in microgravity environments.
2. Economic: The high initial costs of asteroid mining missions and the long payback periods present significant financial hurdles.
3. Legal and Regulatory: The lack of clear international regulations on space resource utilization creates uncertainty for potential investors.
4. Logistical: Identifying suitable asteroids, intercepting them, and returning resources to Earth or space-based facilities present complex logistical challenges.

Leading Companies in Asteroid Mining:

- AstroForge (https://www.astroforge.io/)

Focus: Developing technology to mine platinum-group metals from asteroids

- Planetary Resources (acquired by ConsenSys) (https://www.planetaryresources.com/)

Focus: Asteroid prospecting and mining technology development

- Deep Space Industries (acquired by Bradford Space) (https://deepspaceindustries.com/)

Focus: Spacecraft technology for asteroid mining and exploration

- TransAstra Corporation (https://www.transastracorp.com/)

Focus: Developing optical mining technology for asteroid resource extraction

- Asteroid Mining Corporation (https://www.asteroidminingcorporation.co.uk/)

Focus: Developing satellites for asteroid prospecting and mining

- iSpace (https://ispace-inc.com/)

Focus: Lunar exploration and resource utilization, with potential applications for asteroid mining

- Moon Express (https://moonexpress.com/)

Focus: Lunar mining technology, potentially adaptable for asteroid mining

- Karman+ (https://karmanplus.com/)

Focus: Near-Earth asteroid resource utilization

While asteroid mining is still in its early stages, these companies are driving innovation in spacecraft design, resource extraction techniques, and space-based manufacturing. As technology advances and costs decrease, asteroid mining could become a critical component of humanity's expansion into space and a potential solution to resource scarcity on Earth.

Directed Energy: Harnessing Power in Space

Directed energy technologies are at the forefront of modern advancements in space applications. These technologies use concentrated electromagnetic energy to achieve various effects, from communications to propulsion and even potential weapons systems. The main types of directed energy technologies being developed for space applications include:

1. High-Energy Lasers (HELs)
High-Energy Lasers (HELs) use focused beams of light to deliver energy to a target. These lasers can generate immense amounts of heat, capable of damaging or destroying targets with precision.
Applications:

- Defense: HELs can be used to intercept and destroy incoming missiles, drones, and other threats. They offer a low-cost-per-shot alternative to traditional munitions.
- Communications: High-energy lasers can enable high-speed, secure communication links between satellites and ground stations.
- Propulsion: HELs can potentially be used for spacecraft propulsion by heating a propellant or providing direct thrust.

2. High-Power Microwaves (HPMs)
High-Power Microwaves (HPMs) emit high-peak-power radio frequency waves that can disrupt or destroy electronic systems. HPMs can deliver continuous or pulsed microwave energy to a target.

Applications:

- Electronic Warfare: HPMs can disable or destroy enemy electronics, including communication systems, radar, and control systems.
- Area Denial: HPMs can be used to create zones where electronic devices are rendered inoperative, providing strategic advantages in military operations.
- Space-Based Power Transmission: HPMs can be used to transmit power from space-based solar arrays to Earth or other spacecraft.

3. Particle Beams

Particle beams use streams of charged or neutral particles accelerated to high velocities to deliver energy to a target. These beams can penetrate deeply into materials, causing damage through thermal energy and radiation.

Applications:

- Propulsion: Particle beams can be used for spacecraft propulsion by providing thrust through the acceleration of particles.
- Defense: Particle beam weapons can damage or destroy targets by irradiating them with high-energy particles.
- Scientific Research: Particle beams are used in various scientific applications, including particle physics experiments and materials testing.

Key Companies:
- Raytheon Technologies
 (https://www.rtx.com/raytheon/what-we-do/integrated-air-and-missile-defense/lasers)
- Raytheon Technologies (https://www.rtx.com/raytheon/what-we-do/advanced-technology/high-power-microwaves)

- Lockheed Martin
 (https://www.lockheedmartin.com/en-us/capabilities/directed-energy.html)
- Northrop Grumman (https://www.northropgrumman.com/what-we-do/air/directed-energy/)
- Boeing (https://www.boeing.com/defense/directed-energy-systems/)
- L3Harris Technologies (https://www.l3harris.com/all-capabilities/directed-energy)

Conclusion

As the technology continues to advance, directed energy systems are likely to play an increasingly important role in future space operations, potentially revolutionizing various aspects of space exploration, communication, and defense.

High-Energy Lasers, High-Power Microwaves, and Particle Beams each offer unique capabilities that can be harnessed for a wide range of applications, from enhancing spacecraft propulsion to providing robust defense mechanisms against emerging threats. The ongoing development and deployment of these technologies by leading companies will shape the future of space operations, making directed energy an integral part of our journey into the cosmos.

The Convergence of Earth Observation and Space Resources

Using Earth observation techniques to identify and characterize near-Earth asteroids

The convergence of Earth observation (EO) technologies and space resource utilization is opening new frontiers in our understanding and exploitation of near-Earth asteroids (NEAs). This section explores how EO techniques are being used to identify and characterize NEAs, the role of advanced data processing and AI, and the implications for space resource assessment.

Earth observation techniques are critical for detecting and characterizing NEAs, which are potential sources of valuable resources such as metals, water, and rare earth elements. These techniques include:

1. Ground-Based Telescopes: Systems like the Catalina Sky Survey, ATLAS, and Pan-STARRS provide continuous monitoring of the sky to detect new NEAs. These telescopes capture multiple images of the same region of the sky, identifying moving objects against a fixed background of stars and galaxies.

2. Space-Based Telescopes: Missions such as NEOWISE and the upcoming NEO Surveyor use infrared technology to detect and characterize asteroids. These space-based observatories can identify both bright and dark asteroids, providing crucial data on their size, composition, and orbits.

3. Spectral Analysis: Hyperspectral imaging technologies, which capture data across a wide range of wavelengths, are used to determine the composition of asteroids. This helps in identifying resource-rich asteroids that contain valuable metals and water.

Advanced Data Processing and AI Applications
The integration of advanced data processing and artificial intelligence (AI) is enhancing our ability to analyze EO data for space resource applications:

1. Machine Learning Algorithms: Tools like the THOR algorithm, developed by the Asteroid Institute and the University of Washington, enable the discovery of new asteroids from archival data. These algorithms can process vast amounts of data quickly and accurately, identifying potential NEAs for further study.

2. AI Foundation Models: Collaborations such as NASA and IBM's HLS Geospatial Foundation Model demonstrate the potential for AI to extract insights from EO data. These models can analyze large datasets to identify patterns and anomalies, aiding in the characterization of NEAs.

3. Data Integration: The European Union's planned digital twin of Earth aims to integrate all available global observations for model development and applications. This type of integration can transcend institutional barriers and be applied to other areas of Earth science, including asteroid detection and characterization

Asteroid Characterization for Resource Potential

EO techniques are being adapted to assess the resource potential of NEAs, focusing on identifying and characterizing asteroids with valuable materials:

1. Spectral Analysis: Hyperspectral imaging can determine the mineral composition of asteroids, identifying those rich in metals and water. Companies like Pixxel are developing advanced hyperspectral imaging capabilities for this purpose.

2. Phase Curve Analysis: This technique involves analyzing the light curves of asteroids to infer their composition. Companies like Karman+ are using this method to identify carbonaceous asteroids, which are rich in water and other valuable resources.

3. In-Situ Resource Utilization (ISRU): EO data informs the development of technologies for extracting and processing resources directly on asteroids. This includes identifying suitable mining sites and assessing the feasibility of resource extraction.

Integration of Earth Observation and Space Mining Technologies

As space mining becomes more feasible, EO technologies are being integrated into asteroid mining strategies:

1. Prospecting Satellites: Companies like AstroForge and Asteroid Mining Corporation are developing specialized satellites to identify and characterize resource-rich asteroids. These satellites use advanced EO techniques to map the surface and composition of asteroids.

2. Space-Based Manufacturing: EO data supports the development of space-based manufacturing processes, enabling the extraction and utilization of asteroid resources in space. This reduces the need to transport materials from Earth, making long-term space missions more sustainable.

International Collaboration and Policy Considerations

The convergence of EO and space resource utilization is fostering international cooperation and raising essential policy questions:

    1. Data Sharing: Initiatives like NASA's Open-Source Science Initiative promote the sharing of EO data and asteroid characterization information.

    2. Regulatory Frameworks: As asteroid mining becomes more realistic, there is a growing need for international agreements on space resource utilization. These frameworks will ensure that space resources are used responsibly and sustainably.

As EO technologies continue to advance and the space resources sector matures, the convergence of these fields will drive innovation and expand our capabilities for space exploration and utilization. This synergy will play a crucial role in humanity's journey to become a multi-planetary species.

Conclusion

The convergence of Earth observation (EO) technologies and space resource utilization represents a significant leap forward in our ability to explore and exploit the resources of near-Earth asteroids (NEAs). By leveraging advanced EO techniques, such as ground-based and space-based telescopes, spectral analysis, and machine learning algorithms, we can more accurately identify and characterize NEAs. This improved understanding not only enhances our planetary defense capabilities but also opens new possibilities for sustainable space exploration and economic development.

The integration of EO data with space mining technologies is paving the way for innovative approaches to resource extraction and utilization in space. Prospecting satellites and in-situ resource utilization (ISRU) technologies are being developed to assess and harness the valuable materials found in asteroids, such as metals, water, and rare earth elements. These advancements promise to reduce the need for Earth-based resources, making long-term space missions more feasible and sustainable. In summary, the convergence of EO and space resource technologies is driving innovation and expanding our capabilities in space exploration.

CHAPTER VIII
THE COMMERCIALIZATION OF SPACE

The commercialization of space is transforming the final frontier into a bustling hub of economic activity. This chapter delves into the burgeoning industries within the space sector, focusing on space tourism and the broader business opportunities that are emerging as private companies take the lead in space exploration and utilization.

Space Tourism: Beyond Sightseeing

Space tourism is rapidly evolving from a futuristic concept to a tangible reality, driven by the ambitions of private companies aiming to make space travel accessible to civilians. This section explores the current state and future prospects of space tourism.

Key Aspects and Considerations:

Suborbital Flights: Companies like Virgin Galactic and Blue Origin are pioneering suborbital space tourism, offering passengers a few minutes of weightlessness and breathtaking views of Earth from the edge of space.
  - Virgin Galactic (https://www.virgingalactic.com)
  - Blue Origin (https://www.blueorigin.com)

Orbital Missions: SpaceX is leading the charge in orbital space tourism, planning missions that will take private individuals around the Earth and even to the Moon.
- SpaceX (https://www.spacex.com)

Commercial Space Stations: Future plans include developing commercial space stations and habitats where tourists can stay for extended periods, offering unique environments for recreation, research, and manufacturing.
  - Axiom Space (https://www.axiomspace.com)

Challenges and Opportunities: The high cost of space travel remains a significant barrier, but advancements in technology and increased competition are expected to drive down prices over time. Regulatory frameworks are also being developed to ensure passenger safety and responsible use of space.

Top Companies in Space Tourism:

- SpaceX (https://www.spacex.com)
- Blue Origin (https://www.blueorigin.com)
- Virgin Galactic (https://www.virgingalactic.com)
- Axiom Space (https://www.axiomspace.com)
- Space Adventures (https://www.spaceadventures.com)

The Business of Space: Economics and Opportunities

The commercialization of space extends beyond tourism, encompassing a wide range of economic activities and opportunities. This section examines the various business ventures and the economic potential of space.

Key Areas of Economic Activity:

Satellite Services: The satellite industry is a cornerstone of the space economy, providing services such as communications, Earth observation, and navigation. Companies like SES, Eutelsat, and Inmarsat are major players in this sector.
  - SES (https://www.ses.com)
  - Eutelsat (https://www.eutelsat.com)
  - Inmarsat (https://www.inmarsat.com)

Launch Services: The demand for satellite launches has spurred the growth of launch service providers. SpaceX, Rocket Lab, and Blue Origin are leading the way with innovative and cost-effective launch solutions.

Space Manufacturing and Resource Utilization: Companies are exploring the potential of manufacturing in space and utilizing space resources. This includes in-space manufacturing of components and the extraction of resources from asteroids and the Moon.

Space Habitats and Infrastructure: The development of space habitats and infrastructure is crucial for long-term human presence in space. Companies like Orbital Assembly Corporation are working on creating sustainable living environments in space.

Top Companies in the Business of Space:

- SpaceX (https://www.spacex.com)
- Blue Origin (https://www.blueorigin.com)
- Virgin Galactic (https://www.virgingalactic.com)
- Rocket Lab (https://www.rocketlabusa.com)
- SES (https://www.ses.com)
- Eutelsat (https://www.eutelsat.com)
- Inmarsat (https://www.inmarsat.com)
- Made In Space (https://www.madeinspace.us)
- AstroForge (https://www.astroforge.io)
- Bigelow Aerospace (https://www.bigelowaerospace.com)
- Orbital Assembly Corporation (https://www.orbitalassembly.com)

Public-Private Partnerships:

Public-private partnerships are playing a crucial role in the commercialization of space. Governments and space agencies are increasingly collaborating with private companies to leverage their expertise and resources. This collaboration is essential for large-scale projects such as lunar bases and Mars missions.

- NASA's Commercial Crew Program (https://www.nasa.gov/exploration/commercial/crew)

Space Law and Policy:

As commercial activities in space increase, the need for clear legal and regulatory frameworks becomes more pressing. This section explores the current state of space law and the efforts to develop policies that ensure the responsible and sustainable use of space.
- United Nations Office for Outer Space Affairs (UNOOSA) (https://www.unoosa.org)

Environmental Impact and Sustainability:

The environmental impact of space activities, particularly rocket emissions and space debris, is a growing concern. Companies and organizations are working on solutions to minimize these impacts and promote sustainable practices in space exploration and utilization.
- Astroscale (https://astroscale.com)

SpaceX and Blue Origin are taking several approaches to make space travel more sustainable:

1. Reusable Rockets: Both companies have developed reusable launch vehicles to significantly reduce costs and waste. SpaceX's Falcon 9 and Falcon Heavy, and Blue Origin's New Shepard and planned New Glenn rockets are designed for multiple launches.

2. Cleaner Fuel: Blue Origin uses liquid oxygen and hydrogen to fuel New Shepard, and liquefied natural gas for New Glenn. These fuels produce fewer emissions compared to traditional rocket propellants.

3. Advanced Materials and Manufacturing: Both companies are investing in advanced materials and manufacturing techniques to create lighter, more efficient spacecraft components.

4. Debris Mitigation: SpaceX has committed to responsible behavior in orbit, proactively de-orbiting satellites that may pose future risks of becoming space debris.

5. Sustainable Supply Chains: Blue Origin is implementing sustainable practices in its supply chain, focusing on reducing, reusing, and recycling materials.

SPACEPOSIUM

6. Earth Observation and Climate Monitoring: While not directly related to space travel, both companies' satellite technologies contribute to Earth observation and climate monitoring, indirectly supporting sustainability efforts on Earth.

7. Long-term Vision: Both companies have long-term visions that include space resource utilization and off-Earth manufacturing, which could reduce the need for Earth-based resources in future space missions.

While these efforts are significant steps towards sustainability, challenges remain, including the overall carbon footprint of increased launch activities and the potential environmental impacts of large satellite constellations. Both companies continue to research and develop technologies to address these concerns and improve the sustainability of space travel.

Conclusion

The commercialization of space is transforming the aerospace industry and revolutionizing education, inspiring the next generation, and creating unprecedented opportunities for global collaboration. Initiatives like Spaceposium are at the forefront of this transformation, bridging the gap between cutting-edge space technology and classrooms around the world. By upgrading school audiovisual equipment and providing space-related educational resources, Spaceposium is democratizing access to space education and inspiring students across the globe.

As private companies push the boundaries of space exploration and utilization, they are fostering a new era of educational engagement and creating new career paths and economic opportunities. The space economy is rapidly evolving beyond traditional government-led programs into a diverse ecosystem of public-private partnerships, commercial ventures, and educational initiatives. This shift is driving technological advancements and promises to be a significant catalyst for global economic growth, scientific discovery, and innovation.

The vision of a thriving space economy that benefits all of humanity is becoming increasingly tangible. From space tourism to satellite services, and from manufacturing in microgravity to resource utilization on other celestial bodies, the possibilities are vast. However, the true potential of this new frontier will only be realized if we continue to invest in education, foster international cooperation, and ensure equitable access to space-related opportunities. By combining education, innovation, and inspiration, initiatives like Spaceposium are laying the groundwork for a future where space is not just a realm of exploration, but an integral part of our global economy and society.

CHAPTER IX
THE CURRENT SPACE INDUSTRY LANDSCAPE

SPACEPOSIUM

This chapter will explore how established aerospace companies, new private entrants, and international space agencies are navigating this rapidly evolving landscape. We'll examine the economic opportunities, technological advancements, and challenges facing the industry as it moves towards a more commercially driven future. From the rise of space tourism to the potential of space manufacturing and resource utilization, we'll provide a comprehensive overview of the current state and future prospects of the space industry.

The space industry is undergoing a transformative era marked by significant advancements, increased private investment, and the emergence of new players. This chapter provides an overview of the current landscape, highlighting established aerospace giants, pioneering private companies, innovative startups, key international space agencies, and the burgeoning commercial space economy.

The global space sector is experiencing rapid growth, with forecasts suggesting it could be worth around $1.1 trillion by 2030, implying an annual growth rate of approximately 11% annually This growth is driven by a shift from traditional government-led endeavors to a more commercially driven "New Space" sector, characterized by private companies and startups driving innovation and exploration.

Key areas of innovation include reusable rockets, small satellites, space tourism, and in-space manufacturing, which are reshaping the industry landscape. The U.S. private sector space workforce jumped 4.8% in 2023, with a strong employment forecast promising more work ahead, especially in the growing commercial space market.
The industry is witnessing a transition towards cost-effective and scalable technologies. Small satellites have become a leading trend, with their miniaturization allowing for cost-effective designs and mass production. The small satellite market alone is estimated to be valued at USD 166.40 billion in 2024 and is expected to reach USD 260.56 billion by 2029.

Giants of the Space Race: Established Aerospace Companies

Established aerospace companies have long been the backbone of space exploration and defense. These giants continue to play a crucial role in advancing space technology and infrastructure.

- Boeing (https://www.boeing.com/space/)
Boeing has been a key player in space exploration, contributing to projects like the International Space Station (ISS) and developing the CST-100 Starliner spacecraft for crewed missions.

- Lockheed Martin (https://www.lockheedmartin.com/en-us/capabilities/space.html)
Lockheed Martin is involved in various space missions, including the Mars rovers and the Orion spacecraft, which is designed for deep space exploration.

- Northrop Grumman (https://www.northropgrumman.com/what-we-do/space/)Northrop Grumman provides critical technologies for space exploration, including the Cygnus spacecraft for ISS resupply missions and the James Webb Space Telescope.

New Space Pioneers: Private Space Companies

Private companies are revolutionizing the space industry with innovative technologies and ambitious goals, making space more accessible and affordable.

- SpaceX (https://www.spacex.com)Founded by Elon Musk, SpaceX is known for its reusable Falcon rockets and the Starship spacecraft, which aims to enable human missions to Mars.
- Blue Origin (https://www.blueorigin.com)Founded by Jeff Bezos, Blue Origin focuses on reusable rocket technology with its New Shepard and New Glenn rockets, aiming to lower the cost of space travel.
- Virgin Galactic (https://www.virgingalactic.com)Virgin Galactic is pioneering suborbital space tourism with its SpaceShipTwo, offering civilians the opportunity to experience space travel.

Innovative Space Startups

Innovative startups are driving the next wave of space exploration with cutting-edge technologies and novel business models.

- Rocket Lab (https://www.rocketlabusa.com)Rocket Lab specializes in small satellite launches with its Electron rocket, providing dedicated and cost-effective access to space.
- Relativity Space (https://www.relativityspace.com)Relativity Space is revolutionizing rocket manufacturing with 3D printing technology, aiming to build rockets more quickly and efficiently.
- Astroscale (https://astroscale.com)Astroscale focuses on space debris removal and sustainable space operations, developing technologies to mitigate the growing problem of space junk.

International Space Agencies and Missions

International space agencies continue to lead in scientific research, exploration missions, and international collaboration.

- NASA (https://www.nasa.gov)NASA is the leading space agency in the United States, responsible for groundbreaking missions like the Mars rovers, the Artemis program, and the Hubble Space Telescope.
- ESA (European Space Agency) (https://www.esa.int)ESA conducts a wide range of missions, from Earth observation to planetary exploration, and collaborates with other space agencies on international projects.
- JAXA (Japan Aerospace Exploration Agency) (https://www.jaxa.jp)JAXA is known for its contributions to space science, including the Hayabusa asteroid missions and participation in the ISS.

- Roscosmos (https://www.roscosmos.ru)Roscosmos, the Russian space agency, has a long history of space exploration, including human spaceflight and interplanetary missions.
- ISRO (Indian Space Research Organisation) (https://www.isro.gov.in)ISRO has made significant strides in space technology, with notable missions like the Mars Orbiter Mission (Mangalyaan) and the Chandrayaan lunar missions.

The Commercial Space Economy: Satellites, Tourism, and Beyond

The commercial space economy is expanding rapidly, driven by advancements in satellite technology, space tourism, and other innovative applications. This section explores the key components of this burgeoning sector, highlighting the significant trends and the companies leading the charge.

Satellite Technology

Satellite technology has seen remarkable advancements, transforming the space industry and enabling a wide range of applications:

1. Miniaturization and Small Satellites: The development of small satellites, such as CubeSats, has revolutionized the industry by making space more accessible and affordable. These compact satellites are used for various purposes, including Earth observation, communications, and scientific research.Planet Labs
2. Large Satellite Constellations: Companies are deploying large constellations of satellites in low Earth orbit (LEO) to provide global internet coverage and other services.SpaceX's Starlink
3. Advanced Propulsion and In-Orbit Servicing: Innovations in propulsion systems and in-orbit servicing are enhancing satellite capabilities and extending their operational lifetimes.Northrop Grumman
4. Data Processing and AI: The integration of artificial intelligence and advanced data processing techniques is improving the efficiency and effectiveness of satellite operations.BlackSky

Space Tourism

As humanity stands on the precipice of a new era, the once-distant dream of space tourism is rapidly becoming a breathtaking reality. The cosmos, long the domain of elite astronauts and scientific missions, is now opening its celestial gates to civilians, promising experiences that will redefine our perspective of Earth and our place in the universe.

From suborbital joyrides to luxurious stays in orbiting hotels, a new industry is taking shape, one that promises to transform not just how we view space, but how we view ourselves as a species. Welcome to the dawn of space tourism, where the sky is no longer the limit, and the stars are now within reach.

Here are some of the key players in this burgeoning industry:

Virgin Galactic (https://www.virgingalactic.com)
Virgin Galactic offers suborbital spaceflights using its air-launched SpaceShipTwo spaceplane. Passengers experience several minutes of weightlessness and view the curvature of Earth. Tickets are priced at $450,000 per person.

Blue Origin (https://www.blueorigin.com)
Jeff Bezos' Blue Origin provides suborbital flights on its New Shepard reusable rocket. The capsule accommodates six passengers and travels to the edge of space in an 11-minute journey, including about four minutes of weightlessness.

• Blue Origin's astronaut training program is a concise yet comprehensive process designed to prepare individuals for suborbital spaceflight on the New Shepard vehicle. The training spans 14 hours over two days and focuses on essential knowledge and skills needed for a safe and enjoyable space experience. Participants receive classroom instruction on mission profiles and spacecraft systems, engage in demonstrations, and practice in a training capsule. The program covers vehicle familiarization, safety procedures, zero-G protocols, ingress and egress techniques, and communication with ground control.

• The training culminates in mission rehearsals covering five different scenarios and a final exam to ensure readiness. Blue Origin evaluates astronauts on various requirements, including physical abilities such as climbing the launch tower and moving within the capsule. While the program is designed to be accessible for non-professional astronauts, it still maintains compliance with Federal Aviation Administration (FAA) regulations. This streamlined approach reflects the automated nature of the New Shepard system, which requires no piloting from passengers.

SpaceX (https://www.spacex.com)
Elon Musk's SpaceX offers more ambitious orbital flights using its Crew Dragon spacecraft. The company has already conducted private orbital missions and plans future trips around the Moon.

Axiom Space (https://www.axiomspace.com)
Axiom Space focuses on orbital tourism, offering private missions to the International Space Station. The company is also developing its own commercial space station, Axiom Station.

Space Perspective (https://spaceperspective.com)
Space Perspective is developing a unique approach using a high-altitude balloon called Spaceship Neptune. The pressurized capsule will carry eight passengers to an altitude of 100,000 feet for a six-hour journey, including the Space Spa, a luxurious bathroom facility.

World View (https://www.worldview.space)
World View plans to offer high-altitude balloon flights to 100,000 feet, providing a five-day experience including a ground tour and a 6-8 hour balloon journey.

Orbital Assembly Corporation (https://orbitalassembly.com)
Orbital Assembly is developing rotating space stations that will provide artificial gravity. Their projects include Voyager Station for 400 guests and the smaller Pioneer Station for 28 guests.

Space Spa (part of Space Perspective's offering)
The Space Spa is a unique feature of Spaceship Neptune. It offers a luxurious bathroom experience during the six-hour spaceflight. It's designed to surpass first-class airplane facilities and provide a serene sanctuary with spectacular views.

These pioneering companies are not just selling tickets to space; they're offering a transformative experience that promises to change how we perceive our planet and our role in the cosmos. As the industry evolves, we can expect more innovations, increased accessibility, and perhaps even the first actual space hotels, marking the beginning of humanity's expansion beyond Earth.

Conclusion

The current space industry landscape is characterized by a dynamic mix of established aerospace giants, pioneering private companies, innovative startups, and international space agencies. This diverse ecosystem is driving the rapid commercialization of space, making it more accessible and sustainable than ever before. The industry is experiencing unprecedented growth, with forecasts suggesting it could be worth around USD 1.1 trillion by 2030, implying an annual growth rate of approximately 11% annually.

As the industry continues to evolve, collaboration between these entities will be crucial in overcoming challenges and unlocking new opportunities. Public-private partnerships, such as NASA's Commercial Crew Program, demonstrate the power of combining government resources with private-sector innovation. These collaborations are not only accelerating technological advancements but also reducing costs and increasing efficiency in space operations.

Key challenges facing the industry include the need for sustainable practices in space, managing the increasing problem of space debris, and addressing the potential militarization of space. Additionally, the industry must navigate complex regulatory environments and work towards establishing clear international frameworks for space resource utilization and traffic management.

Despite these challenges, the opportunities for exploration, economic growth, and technological advancement are immense. The rise of small satellites and advanced manufacturing techniques opens new possibilities for Earth observation, communications, and scientific research. Space tourism is becoming a reality, with companies like Virgin Galactic and Blue Origin leading the way. Meanwhile, ambitious plans for lunar bases and Mars missions are pushing the boundaries of human exploration.

The future of space is bright, with endless possibilities for innovation and discovery. As we look ahead, we can expect to see continued advancements in areas such as artificial intelligence in space operations, sustainable space technologies, and in-situ resource utilization. The commercialization of space creates new industries and job opportunities and promising solutions to some of Earth's most pressing challenges, from climate change monitoring to global communications.

In conclusion, the space industry stands at the cusp of a new era. The convergence of technological innovation increased private investment, and international cooperation set the stage for a future where space plays an integral role in our daily lives and the global economy. As we continue to push the boundaries of what's possible, the space industry will undoubtedly be a driving force in shaping the future of humanity, both on Earth and beyond.

CHAPTER X
COLLABORATION AND EDUCATION

International cooperation in space exploration has become a cornerstone of progress, enabling nations to pool resources, share expertise, and achieve common goals that would be challenging to accomplish individually. This collaborative approach advances scientific knowledge and technological capabilities, fosters diplomatic ties, and promotes peaceful uses of outer space.

The International Space Station (ISS) is a prime example of successful international cooperation. Launched in 1998, the ISS is supported by a partnership of space agencies from the United States (NASA), Russia (Roscosmos), Europe (ESA), Japan (JAXA), and Canada (CSA). This collaboration has allowed for shared resources, expertise, and costs, leading to significant scientific discoveries and technological advancements. The ISS has provided a platform for over 3,000 scientific experiments conducted by researchers from more than 100 countries, demonstrating the immense benefits of international cooperation.

The Artemis Accords, introduced in October 2020, represent another significant step towards international collaboration in space. These accords establish a set of principles to guide space cooperation among countries participating in NASA's Artemis program, which aims to return humans to the Moon and eventually send astronauts to Mars. The Artemis Accords promote transparency, interoperability, and peaceful exploration, ensuring that space activities benefit all of humanity.

Joint missions and collaborative projects further exemplify the benefits of international cooperation. For instance, the BepiColombo mission to Mercury, a collaboration between ESA and JAXA, and the ExoMars mission, a partnership between ESA and Roscosmos, highlight how pooling resources and expertise can lead to groundbreaking scientific achievements. Additionally, initiatives like the Combined Space Operations (CSpO) initiative and the AUKUS trilateral security partnership demonstrate how international cooperation can enhance space security and operational capabilities.

Despite the numerous benefits, international cooperation in space also presents challenges, including political and diplomatic complexities, differing levels of technological capabilities, and varying financial investments. However, nations can overcome these challenges through open dialogue and a commitment to finding common ground and forging strong partnerships. The pursuit of shared goals in space exploration fosters a sense of unity and serves as a testament to our collective ability to work towards common objectives, transcending earthly boundaries.

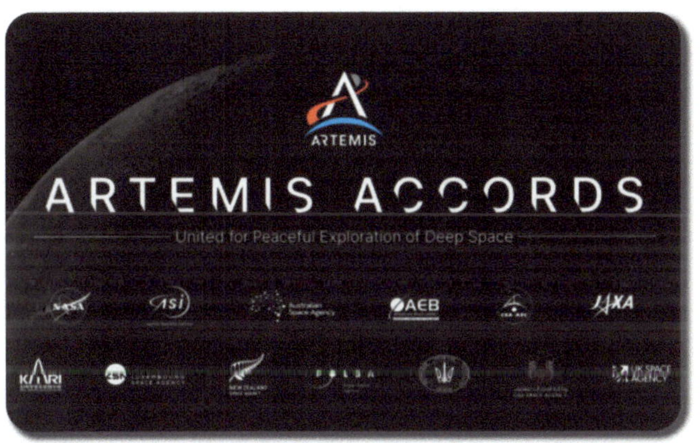

LEARNING IN ORBIT: THE EVOLUTION OF SPACE EDUCATION

Space education has undergone a significant transformation in recent years, driven by technological advancements and a growing emphasis on STEM fields. This section explores the innovative approaches and programs that are shaping the future of space education. From interactive exhibits and virtual reality to citizen science projects and international collaborations, these initiatives inspire the next generation of scientists, engineers, and explorers.

Interactive Exhibits and Virtual Reality

Museums and space centers are leveraging cutting-edge technology to create immersive learning experiences. For example:

- **Kennedy Space Center Visitor Complex** (https://www.kennedyspacecenter.com) offers interactive exhibits that simulate space exploration experiences.
- **Smithsonian National Air and Space Museum** (https://airandspace.si.edu) provides virtual reality (VR) experiences that allow students to take virtual tours of planets and moons, enhancing their understanding of space environments.

Citizen Science Projects

Space agencies are engaging the public in scientific research through citizen science initiatives:

- **NASA's Citizen Science Projects** (https://science.nasa.gov/citizenscience) enable the public to contribute data and observations using their own equipment.
- These projects foster a sense of participation in space exploration and help collect valuable scientific data.

Social Media and Online Platforms

Space agencies and organizations are using social media to educate and inspire:

- NASA's presence on platforms like Twitter, Instagram, and TikTok provides updates on missions and behind-the-scenes glimpses into space operations.
- These efforts help make space science more accessible and engaging for younger audiences.

SPACEPOSIUM

STEM Education Programs

Specialized programs are being developed to integrate space themes into STEM education:

- NASA's HUNCH (High Schools United with NASA to Create Hardware) (https://www.nasa.gov/audience/foreducators/hunch/index.html) program provides students hands-on experience designing and building hardware for space missions.
- The Space Prize Foundation's curriculum (https://www.spaceprize.org) prepares students for the growing space economy and humanity's multi-planet future.

International Space Station as an Educational Platform

The ISS serves as a unique classroom for students around the world:

- ISS National Laboratory educational programs (https://www.issnationallab.org/stem/) offer students opportunities to conduct microgravity experiments.
- Live video connections with astronauts inspire students and provide real-time insights into life in space.

Diversity and Inclusion Efforts

There's a growing focus on increasing representation in space-related fields:
- Programs aim to encourage participation from underrepresented groups in STEM fields.
- NASA's Artemis program (https://www.nasa.gov/specials/artemis/) aims to land the first woman and the next man on the Moon, promoting gender equality in space exploration.

Global Collaboration in Education

International partnerships are enhancing space education:
- Programs like the Vera C. Rubin Observatory's international partnerships (https://www.lsst.org) enable students and researchers from different countries to work together on cutting-edge astronomical research.
- These collaborations promote cultural exchange and broaden the global knowledge base in space science.

Augmented Reality in Space Education

AR technologies are being used to create interactive learning experiences:
- ESA's VR resources (https://www.esa.int/Education/VR_resources) and NASA's VR/AR resources (https://www.nasa.gov/stem/virtual-reality) allow students to explore space concepts in a hands-on, visual manner.
- These tools help make complex space concepts more accessible and engaging for learners of all ages.

Key Organizations Leading Space Education Initiatives

- NASA (https://www.nasa.gov/stem/)
- European Space Agency (ESA) Education Office (https://www.esa.int/Education)
- Space Foundation (https://www.discoverspace.org/education/)
- The Planetary Society (https://www.planetary.org/outreach)
- Students for the Exploration and Development of Space (SEDS) (https://seds.org/)

Here is a list of significant space education venues, including Space Center Houston and the Virginia Air and Space Museum, along with their URLs for student learning and access by STEM educators:

1. Space Center Houston
URL: https://spacecenter.org/education-programs/
Offers a variety of educational programs, including distance learning, STEM challenges, and educator resources.

2. Virginia Air and Space Center
URL: https://vasc.org/education/
Provides STEM education programs, workshops, and online resources for educators.

3. Smithsonian National Air and Space Museum
URL: https://airandspace.si.edu/learn
Features extensive learning resources, including hands-on activities, videos, and educator professional development.

4. NASA Jet Propulsion Laboratory Education
URL: https://www.jpl.nasa.gov/edu/
Offers a wide range of STEM education resources, lessons, and activities related to space exploration.

5. Kennedy Space Center Visitor Complex
URL: https://www.kennedyspacecenter.com/education
Provides educational programs, including virtual field trips and STEM activities.

6. Adler Planetarium
URL: https://www.adlerplanetarium.org/learn/educators/
Offers educator resources, professional development, and online learning programs.
7. Challenger Center
URL: https://challenger.org/educators/
Provides hands-on STEM education programs and resources for educators.
8. U.S. Space & Rocket Center
URL: https://www.rocketcenter.com/education
Offers various educational programs, including Space Camp and educator resources.
9. Chabot Space & Science Center
URL: https://chabotspace.org/education/
Provides STEM education programs, workshops, and resources for educators.
10. Griffith Observatory
URL: https://griffithobservatory.org/education/
Offers educational resources, including online programs and educator workshops.
11. The Planetary Science Institute
URL: https://www.psi.edu/epo/
Provides educational resources and programs related to planetary science.
12. Space Foundation Discovery Center
URL: https://discoverspace.org/education/
Offers various STEM education programs and resources for educators.

These venues provide a wealth of resources for STEM educators, including lesson plans, virtual field trips, hands-on activities, and professional development opportunities. Many of these institutions have adapted their offerings to include more online and distance learning options, making their resources accessible to educators and students worldwide.

NASA's Next-Gen STEM program engages K-12 students in several key ways:
1. Hands-on Activities and Experiments:
   The program offers a wide array of hands-on activities and experiments designed to engage students in STEM topics related to NASA's missions.
   https://www.nasa.gov/stem-ed-resources/hands-on-activities.html
2. Curriculum Resources:
   NASA develops and provides curriculum resources aligned with national science, math, and technology standards.
   https://www.nasa.gov/stem-ed-resources/curriculum-resources.html
3. Challenges and Competitions:
   NASA offers various challenges and competitions for K-12 students, such as the TechRise Student Challenge where student teams in grades 6-12 design and build experiments for suborbital flight tests.
   https://www.nasa.gov/stem/techrise/
4. Virtual and Augmented Reality Experiences:

The program utilizes VR and AR technologies to create immersive learning experiences about space and NASA missions.

https://www.nasa.gov/stem/virtual-reality/

5. Educator Support:

NASA CONNECTS provides a community of practice for K-12 teachers and informal educators, offering access to NASA resources and experts.

https://www.nasa.gov/stem/nasa-connects/

6. Diverse Learning Opportunities:

Next Gen STEM aims to reach students through various interests, including writing contests and collaborations with arts organizations to engage students who may not initially see themselves in STEM.

https://www.nasa.gov/stem-ed-resources/stem-engagement/

7. Partnerships with Museums and Informal Education Institutions:

NASA partners with museums, science centers, and other institutions to bring space science to communities across the nation.

https://www.nasa.gov/stem/team-ii-program/

8. Focus on Underrepresented Groups:

The program has a specific focus on increasing diversity and inclusion in STEM fields, with efforts to reach underrepresented and underserved students.

https://www.nasa.gov/stem/murep/

9. Real-world Connections:

Next Gen STEM emphasizes connecting students to real NASA missions and projects, like the Artemis program, to make learning more relevant and exciting.

https://www.nasa.gov/specials/artemis/

10. Multimedia Resources:

The program offers various multimedia resources, including videos, interactive websites, and social media content to engage students in different ways.

https://www.nasa.gov/stem-ed-resources/multimedia/

These resources showcase NASA's commitment to engaging K-12 students in STEM education through diverse and innovative approaches.

Conclusion

The landscape of space exploration and education is undergoing a profound transformation, driven by international cooperation, technological advancements, and innovative educational approaches. As we've seen, global partnerships exemplified by initiatives like the International Space Station and the Artemis Accords are pushing the boundaries of what we can achieve in space. These collaborations not only advance scientific knowledge and technological capabilities but also foster diplomatic ties and promote peaceful cooperation among nations.

Simultaneously, space education has evolved dramatically, leveraging cutting-edge technologies and diverse platforms to inspire and educate the next generation of space explorers. Students now have unprecedented access to space-related learning opportunities, from virtual reality experiences and citizen science projects to hands-on experiments on the ISS. Programs like NASA's Next-Gen STEM are making space science more accessible, interactive, and inclusive, helping to build a diverse and skilled workforce ready to tackle the challenges of future space exploration.

The integration of advanced technologies, hands-on learning experiences, and global collaborations ensures that students are not only educated but also inspired to pursue careers in space science and technology. This holistic approach to space education is crucial for fostering a culture of innovation and exploration that will drive humanity's future in space.

As we look to the future, the convergence of international collaboration and innovative education will be crucial in addressing the challenges of space exploration and utilizing space resources for the benefit of humanity. By fostering a global, diverse, and well-educated workforce, we are laying the foundation for a future where space exploration continues to push the boundaries of human knowledge and capability.

The journey to the stars is no longer the domain of a select few nations or individuals. Instead, it has become a collaborative global effort that inspires and engages people of all ages and backgrounds. As we continue to explore the cosmos, the lessons learned and the technologies developed will not only advance our understanding of the universe but also contribute to solving pressing challenges on Earth.

The future of space exploration is intrinsically linked to our ability to collaborate across borders and inspire the next generation. Through continued investment in international partnerships and innovative education programs, we are ensuring that the spirit of exploration and discovery will continue to thrive, propelling humanity towards a future among the stars.

CHAPTER XI
ETHICAL AND ENVIRONMENTAL CONSIDERATIONS

SPACE LAW: LEGAL AND ETHICAL CHALLENGES

Current Legal Framework
The Outer Space Treaty of 1967 forms the basis of international space law, establishing principles like the peaceful use of space and non-appropriation of celestial bodies. National space laws complement this framework, addressing commercial space activities and liability issues. As space activities evolve, new legal challenges emerge, particularly around resource extraction and space tourism.

Ethical Considerations in Space Exploration

Ethical debates in space exploration center on the value and preservation of space environments. Questions arise about our moral obligations to limit activities on celestial bodies and the ethical treatment of potential extraterrestrial life. These considerations balance scientific curiosity with responsible stewardship of the cosmos.

Challenges in Long-Duration Space Flight

Long-duration space flights pose unique ethical challenges, including the justification for exposing humans to extended periods of isolation and physical risks. Considerations include the psychological impact on astronauts and the long-term health effects of prolonged exposure to the space environment.

Commercialization and Privatization of Space

The increasing role of private companies in space exploration raises ethical questions about the balance between scientific exploration and commercial interests. Issues include the equitable distribution of space resources and the potential for corporate interests to influence space policy and exploration priorities.

International Cooperation and Conflict Resolution

Space exploration requires unprecedented levels of international cooperation, raising questions about equitable participation and benefit-sharing. Potential geopolitical tensions in space necessitate the development of robust conflict resolution mechanisms and shared governance structures.

Protecting the Cosmos: Environmental Impact of Space Activities

Space Debris Mitigation

The proliferation of space debris poses significant risks to operational satellites and future space missions. International efforts focus on developing guidelines and technologies for debris mitigation and removal, balancing the need for space utilization with long-term sustainability.

Environmental Impact of Rocket Launches

Rocket launches have both immediate and long-term environmental impacts, including emissions affecting Earth's atmosphere and potential damage to the ozone layer. Research into more environmentally friendly propulsion systems and launch methods is ongoing to minimize these effects.

Planetary Protection

Planetary protection policies aim to prevent biological contamination of other celestial bodies and protect Earth from potential extraterrestrial contamination. These policies are crucial for preserving the scientific integrity of space exploration and safeguarding both Earth and other planets.

Sustainable Space Exploration

Sustainable space exploration focuses on developing technologies and practices that minimize environmental impact and ensure long-term viability. This includes in-situ resource utilization (ISRU) for long-term missions and developing closed-loop life support systems.

Climate Change Monitoring from Space

Satellites play a crucial role in monitoring and understanding climate change on Earth. Space-based observations provide invaluable data for climate science, helping to track global environmental changes and inform policy decisions.

Light Pollution and Astronomical Observations

The increasing number of satellites, particularly large constellations, raises concerns about light pollution and interference with ground-based astronomical observations. Efforts are underway to mitigate these effects through satellite design and operational practices.

THE OUTER SPACE TREATY

The key principles outlined in the Outer Space Treaty include:

1. The exploration and use of outer space shall be carried out for the benefit and in the interests of all countries and shall be the province of all mankind.
2. Outer space shall be free for exploration and use by all states without discrimination.
3. Outer space is not subject to national appropriation by claim of sovereignty, by means of use or occupation, or by any other means.
4. States shall not place nuclear weapons or other weapons of mass destruction in orbit or on celestial bodies or station them in outer space in any other manner.
5. The Moon and other celestial bodies shall be used exclusively for peaceful purposes. This prohibits their use for testing weapons of any kind, conducting military maneuvers, or establishing military bases, installations, and fortifications.
6. Astronauts shall be regarded as envoys of mankind.
7. States shall be responsible for national space activities whether carried out by governmental or non-governmental entities.
8. States shall be liable for damage caused by their space objects.
9. States shall avoid harmful contamination of space and celestial bodies.
10. States shall promote international cooperation in the peaceful exploration and use of outer space.
11. All stations, installations, equipment and space vehicles on celestial bodies shall be open to representatives of other States Parties to the Treaty on a basis of reciprocity.

These principles form the foundation of international space law and guide the peaceful exploration and use of outer space by all nations.

Conclusion

As humanity ventures further into space, this chapter's ethical and environmental considerations become increasingly critical. The Outer Space Treaty of 1967 provides a foundational legal framework, emphasizing the peaceful use of space, the non-appropriation of celestial bodies, and the responsibility of states for their space activities. However, new legal challenges emerge as space activities evolve, particularly around resource extraction and space tourism. National space laws and international agreements must adapt to address these emerging issues, ensuring that space exploration remains a collective endeavor for the benefit of all humankind.

Ethical considerations in space exploration are equally important. The preservation of space environments and the ethical treatment of potential extraterrestrial life are paramount. Long-duration space flights pose unique ethical challenges, including psychological and physical risks to astronauts.

Protecting the cosmos from the environmental impact of space activities is another critical concern. The proliferation of space debris poses significant risks to operational satellites and future missions, necessitating international efforts to develop guidelines and technologies for debris mitigation. Rocket launches have both immediate and long-term environmental impacts, prompting research into more environmentally friendly propulsion systems. Planetary protection policies are crucial for preventing biological contamination of celestial bodies and safeguarding Earth from potential extraterrestrial contamination.

Sustainable space exploration focuses on developing technologies and practices that minimize environmental impact and ensure long-term viability. Initiatives like the Space Sustainability Rating (SSR) incentivize responsible space operations by evaluating missions based on their sustainability practices. Satellites play a crucial role in monitoring climate change on Earth, providing invaluable data for climate science and informing policy decisions. However, the increasing number of satellites raises concerns about light pollution and interference with ground-based astronomical observations, necessitating efforts to mitigate these effects.

In conclusion, the ethical and environmental considerations of space exploration require ongoing attention and action. By adhering to the principles outlined in the Outer Space Treaty and fostering international cooperation, we can ensure that space exploration is conducted responsibly and sustainably. Balancing the benefits of space activities with the need to protect both Earth and the cosmic environment is essential for the long-term success of humanity's journey into space. Through thoughtful consideration of these ethical and environmental challenges, we can pave the way for a future where space exploration continues to advance human knowledge and capability while preserving the integrity of the cosmos.

CHAPTER XII
THE UNIVERSE'S GRAND DESIGN

The universe is a vast and awe-inspiring expanse, filled with wonders that have captivated human curiosity for millennia. Understanding its fundamental principles is the first step in our journey beyond Earth. This chapter delves into the essential concepts of cosmology, exploring the birth of the universe in the Big Bang, the formation of galaxies, stars, and planets, and the intricate dance of celestial bodies governed by the laws of physics. We will unravel the mysteries of dark matter and dark energy, delve into the nature of black holes, and examine the forces that shape our cosmic environment. By gaining a foundational understanding of the universe, we set the stage for appreciating the extraordinary advancements in space technology and exploration that follow.

The Vastness of Space

Do we really understand what's going on in the "Universe"? In certain areas where there are no street lights, we can see millions of stars in the sky. What do we know about them as individuals? What science has known for decades is that the universe encompasses everything from the smallest particles to the most giant galaxies, filled with a myriad of stars and celestial bodies, all separated by vast, almost unfathomable distances.

Understanding Distances in Light-Years

The standard measurements we use on Earth, such as miles and kilometers, fall short when comprehending the cosmic distances between these celestial bodies. Instead, astronomers use the "light-year" as the benchmark for measuring space. A light-year is the distance that light, traveling at about 299,792 kilometers per second, covers in one Earth year. This comes to approximately 5.88 trillion miles (9.46 trillion kilometers). This immense distance helps us grasp the sheer scale of the Universe.

Visualizing Distances in the Universe

To put the concept of a light-year into perspective, let's consider the distance from Earth to some familiar cosmic objects:

- **The Moon: About 1.28 light-seconds away**

- **The Sun: Roughly eight light-minutes away**

- **The Nearest Star, Proxima Centauri: About 4.24 light-years away**

- **The Center of the Milky Way: Approximately 25,000 light-years away**

- **The Andromeda Galaxy, our nearest galactic neighbor: Roughly 2.5 million light-years away**

These examples illustrate the vastness of space and the variety in distances within our galaxy and beyond. As we explore further, the numbers grow exponentially, highlighting the incredible scale of our universe.

The Role of Light-Years in Understanding the Universe

Using light years to measure distance allows astronomers to map the universe in a way that is comprehensible to the human mind, at least in numerical form. It serves as a reminder of our small place in a cosmos that is both ancient and vast beyond our most accessible comprehension. Every point of light in the night sky is not just a spot of brightness but a distant world, a star, or a galaxy, many of which are millions of light years away from us.

This fundamental understanding of the vastness of space is crucial as it sets the stage for exploring more complex concepts, such as the structure of galaxies, the lifecycle of stars, and the dynamics of cosmic phenomena.

Distances Within Our Solar System

Note: 1 Astronomical Unit (AU) is the average distance from Earth to the Sun, about 93 million miles or 150 million kilometers. For More Information: NASA's Solar System Dynamics Group

(https://ssd.jpl.nasa.gov/?planet_pos)

Important Milestones

Closest Star: The closest star to Earth, Proxima Centauri, is 4.22 light-years away.

Voyager 1: The farthest human-made object, Voyager 1, is more than 14 billion miles from Earth. As of 2021, it takes over 21 hours for a signal from Voyager 1 to reach Earth.

Andromeda Galaxy: The nearest spiral galaxy to the Milky Way, the Andromeda Galaxy, is about 2.537 million light-years away.

Edge of Observable Universe: Given the age of the universe and the speed of light, the furthest we can see is about 46.5 billion light-years away.

Hubble Deep Field: One of the universe's most profound images, capturing galaxies up to 12 billion light-years away.

For More Information: HubbleSite   https://hubblesite.org/

Understanding the sheer scale of the universe helps us grasp the challenges and potential of space exploration. Whether sending robotic explorers to nearby planets or devising telescopes to look billions of years into the past, each step we take is humbling and awe-inspiring, a testament to human ingenuity in the face of incomprehensible vastness.

The Grand Scale of Exploration: Navigating Our Cosmic Journey

Imagine this: if we were to travel at the speed of light, it would still take us over 100,000 years to cross our galaxy, the Milky Way. And ours is just one of billions of galaxies in the observable universe!

Space is so vast that our earthly distance measures—miles or kilometers—become practically irrelevant. Instead, we use the concept of a light-year to encapsulate these incomprehensible distances. A light-year, which amounts to approximately 5.88 trillion miles or 9.46 trillion kilometers, represents the staggering distance light covers in just one year. Considering that light is the fastest thing we know, moving at a speed of 299,792,458 meters per second, this measure clarifies the astronomical distances we speak of.

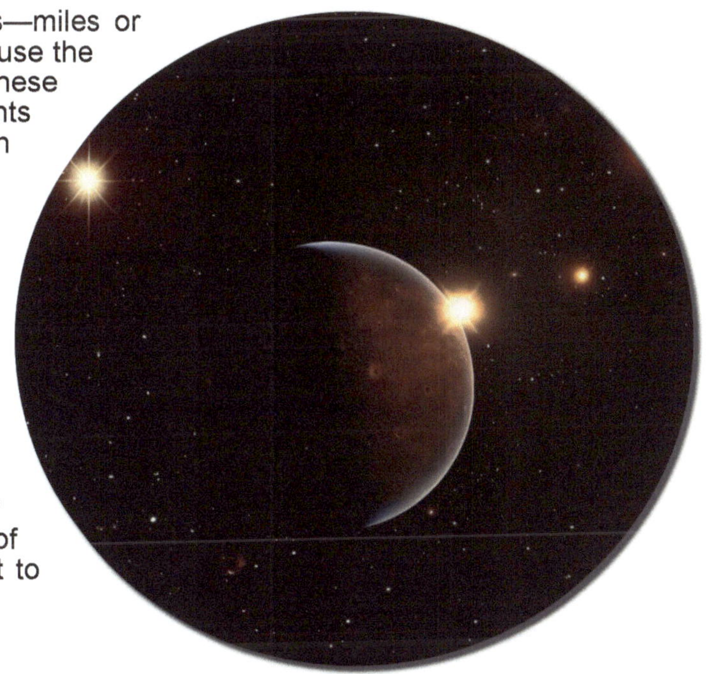

Beyond the confines of the Milky Way, the universe unfurls into even more profound vastness. We live in an observable universe sprinkled with billions of galaxies, each with its unique tapestry of stars, planets, and cosmic phenomena. Some galaxies are similar in size to ours, while others are much larger or smaller. And between these galaxies? Vast stretches of seemingly space that might hold secrets we have yet to discover.

Galaxies: Cities of Stars

Galaxies are colossal systems of stars, star clusters, planetary systems, interstellar gas and dust, and dark matter, all bound together by gravity. They can contain anywhere from a billion to a hundred trillion stars!

The Milky Way, our galactic home, is a barred spiral galaxy with an estimated 100 to 400 billion stars. It spans around 100,000 light-years in diameter. But even with such grandeur, it's a mere speck in the vast cosmic ocean. Some galaxies, like the Andromeda Galaxy, our nearest spiral neighbor, are even more significant than ours. Yet, others, known as dwarf galaxies, are smaller in size but still fascinating in their own right.

Galaxies are monumental assemblies in the vast expanse of space, each a unique congregation of stars, planets, and other celestial wonders. With their intricate networks, these magnificent structures can be considered sprawling star metropolises, each holding a vast population and stories spanning billions of years.

What Comprises a Galaxy?

- Stars: The bright, glowing balls of gas that make up the majority of objects in a galaxy.
- Range: From 1 billion to 100 trillion stars per galaxy.
- Star Clusters: Groups of stars that are gravitationally bound.
- Types: Open and globular clusters.
- Planetary Systems: Planets, moons, and other objects that orbit stars.
- Interstellar Matter: Includes gas and dust floating between stars.
- Nebulae: Birthplaces of stars made of gas and dust.
- Dark Matter: An invisible but pivotal component that holds galaxies together.

Types of Galaxies

Spiral Galaxies: Like our Milky Way, it is characterized by a flat, spinning disk with a central bulge.

Milky Way Facts:

- Estimated 100-400 billion stars.
- About 100,000 light-years in diameter.
- Learn more about the Milky Way

(https://www.nasa.gov/content/goddard/the-milky-way-galaxy)

- Barred Spiral Galaxies: Spiral galaxies with a central bar-shaped structure.
- Elliptical Galaxies: Mostly older stars and not much interstellar matter.
- Dwarf Galaxies: Smaller but dense with stars.
- Role: Often interact with and even "feed" more giant galaxies.

Stars: The Cosmic Lighthouses

Stars are immense celestial bodies primarily composed of hydrogen and helium that undergo nuclear fusion. This process releases vast amounts of energy, lighting up the universe in a dazzling display. Our Sun, a medium-sized star, powers our planetary system and has done so for about 4.6 billion years. It's the gravitational anchor for our solar system and provides the necessary warmth and light for life on Earth.

But stars have life cycles. They are born within the dense, cold pockets of interstellar clouds, live by fusing hydrogen to produce helium, and eventually die. The manner of their end, whether as a white dwarf, neutron star, or black hole, depends on their initial mass.

Stars: Beacons of the Vast Expanse

The night sky, peppered with shimmering points of light, bears testimony to the universe's most brilliant luminaries—stars. Far from being mere static fixtures, these cosmic beacons are dynamic powerhouses that undergo fascinating and intricate life cycles. They play fundamental roles in the cosmic narrative, influencing the formation of galaxies, the creation of elements, and even the possibilities for life itself.

The Nature of Stars

Stars are essentially massive spheres of plasma, primarily made of hydrogen and helium, held together by their own gravity. The heart of a star is its core, where nuclear fusion occurs. This process releases immense energy, which travels outward and eventually radiates into space as light and heat. This light brightens our night sky and has guided humanity for millennia.

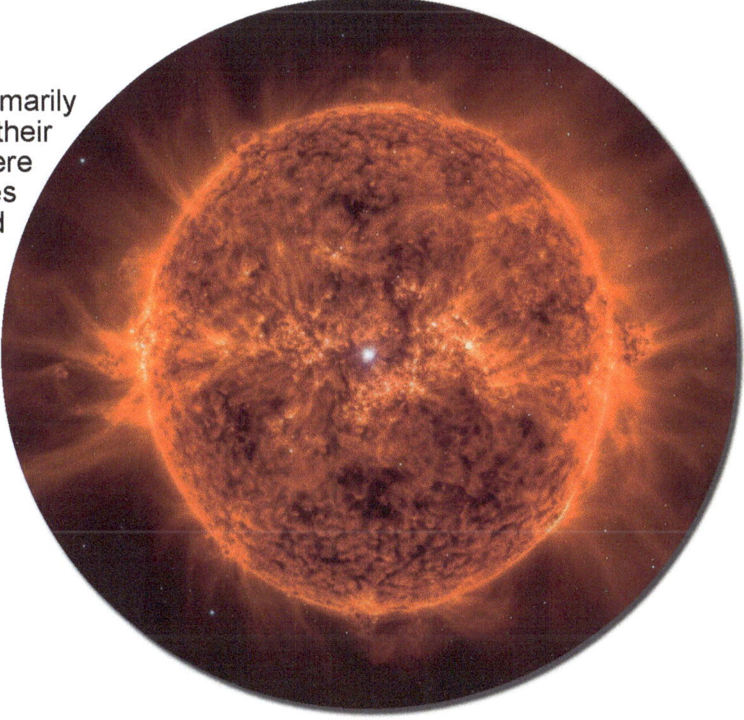

Life Cycle of a Star

The life cycle of a star begins in a nebula, a vast cloud of gas and dust. Under the influence of gravity, parts of this nebula collapse, forming a protostar. As the protostar continues to accumulate mass and heat up, it reaches a point where nuclear fusion ignites, marking its birth as a true star.

The life span of a star depends significantly on its mass:

Low to Medium Mass Stars: Stars like our Sun will fuse hydrogen into helium over billions of years in a stable phase known as the main sequence. After exhausting the hydrogen in their cores, they swell into red giants and eventually shed their outer layers, leaving behind a dense core called a white dwarf.

High Mass Stars: More massive stars have shorter, more tumultuous lives. They burn their nuclear fuel much faster and can end their lives in dramatic supernova explosions, leaving behind neutron stars or black holes.

The Role of Stars in the Cosmos

Stars are the forges of the universe's complex elements. Nuclear fusion creates heavier elements like carbon, nitrogen, and oxygen, essential for life as we know it. When stars end their lives—especially in supernovae— they scatter these elements into space, where they can become part of new stars, planets, and even life forms.

Stars also play a crucial role in galaxies' gravitational architecture. Their collective mass and the gravitational interactions between them help shape galaxies' structure and influence their behavior.

The death of massive stars as supernovae can trigger the formation of new stars, perpetuating the cosmic cycle of stellar evolution. This process enriches galaxies with heavy elements, contributing to the potential habitability of emerging planetary systems.

Observing Stars

When we observe the stars, we are not just looking at points of light but witnessing the ongoing narrative of the universe's evolution. Each star, at each moment in its life, tells a story of cosmic processes that are both violent and beautiful, processes that have profound implications not just for the universe at large but also for our understanding of our origins.

Composition and Fusion: At their core, stars are titanic spheres dominated by hydrogen and helium. These elements aren't just passive components but are actively involved in a cosmic dance. Through nuclear fusion, hydrogen atoms in the star's core are pressed together under immense temperatures and pressures, producing helium and releasing a prodigious amount of energy. This energy radiates outward, manifesting as the radiant light and intense heat that make stars the luminous entities we observe.

SPACEPOSIUM

The Sun—Our Local Beacon: Our very own Sun is central to our understanding of stars. A relatively modest star compared to the gargantuan behemoths scattered across the cosmos, the Sun plays a paramount role in our existence. It has been radiating steadily for about 4.6 billion years, offering light and warmth and setting the rhythm of life on Earth. Acting as the gravitational keystone, it holds our solar system's planets, asteroids, comets, and other celestial bodies in a harmonious ballet.

From Birth to Demise: Stars don't merely flicker into existence; they are birthed in the secluded and frigid recesses of space within nebulous regions known as molecular clouds. These dense areas, rich with gas and dust, collapse under their gravity, forming protostars. Over millions of years, these protostars undergo various stages of development, evolving into fully-fledged stars.

A star's longevity and evolution are profoundly influenced by its mass. While all stars spend most of their lives fusing hydrogen into helium, their eventual fate diverges based on their initial size. Low—to medium-mass stars, like our Sun, will transition into red giants and ultimately shed their outer layers, leaving behind a hot core called a white dwarf. This white dwarf will slowly cool and fade over billions of years.

In contrast, the most massive stars live fast and die young. These colossal entities swell into super giants and meet dramatic ends. When fuel runs out, their cores collapse, leading to spectacular supernova explosions. What remains afterward can be a super dense neutron star or, if massive enough, a mysterious and all-consuming black hole.

Stars, with their mesmerizing luminance and dynamic life cycles, are not just distant points of light; they are the heartbeats of the universe. They have been the subjects of myths and legends, inspired countless dreams, and continue to be at the forefront of our quest to understand the cosmos.

Why Stars Matter

Cosmic Significance: Heartbeats of the universe.
Cultural Impact: Subjects of myths, legends, and modern science.

From their humble origins in interstellar clouds to their diverse end stages, stars are more than just cosmic lighthouses. They're crucial to the very fabric of the universe, lighting up the cosmic sea and acting as gravitational anchors for planetary systems. They've fascinated humanity for millennia and will likely continue to do so as we unravel their myriad mysteries.

Planetary Systems: Islands in the Cosmic Sea

Orbiting around many stars are planets—diverse worlds that range from hot, rocky infernos to cold, gas giants. Our solar system consists of eight primary planets: Mercury, Venus, Earth, Mars, Jupiter, Saturn, Uranus, and Neptune. There are dwarf planets, like Pluto, and countless asteroids, comets, and other small bodies.

Thanks to technological advancements, particularly the Kepler Space Telescope, we've discovered thousands of exoplanets or planets outside our solar system. These discoveries have expanded our understanding of planetary systems and hinted at the possibility that life, in some form, could exist beyond our Earthly confines. In the vastness of the universe, stars act as lighthouses, signaling their presence with brilliance. Yet, circling these luminous giants are planetary systems, the archipelagos of the cosmos, hosting a myriad of worlds, each with its own story and potential.

Diversity of Worlds:

The term "planet" evokes various images: from scorching, metal-molten landscapes to serene, icy realms, from gargantuan gas behemoths that dwarf Earth to modest rocky orbs nestled close to their parent stars. In our solar system alone, the contrast is stark. The inner terrestrial planets - Mercury, Venus, Earth, and Mars - are rocky worlds with solid surfaces. Then, further out, lie the gas giants, Jupiter and Saturn, with their complex systems of rings and moons, followed by the ice giants, Uranus and Neptune, each with its unique atmospheric composition and mysteries.

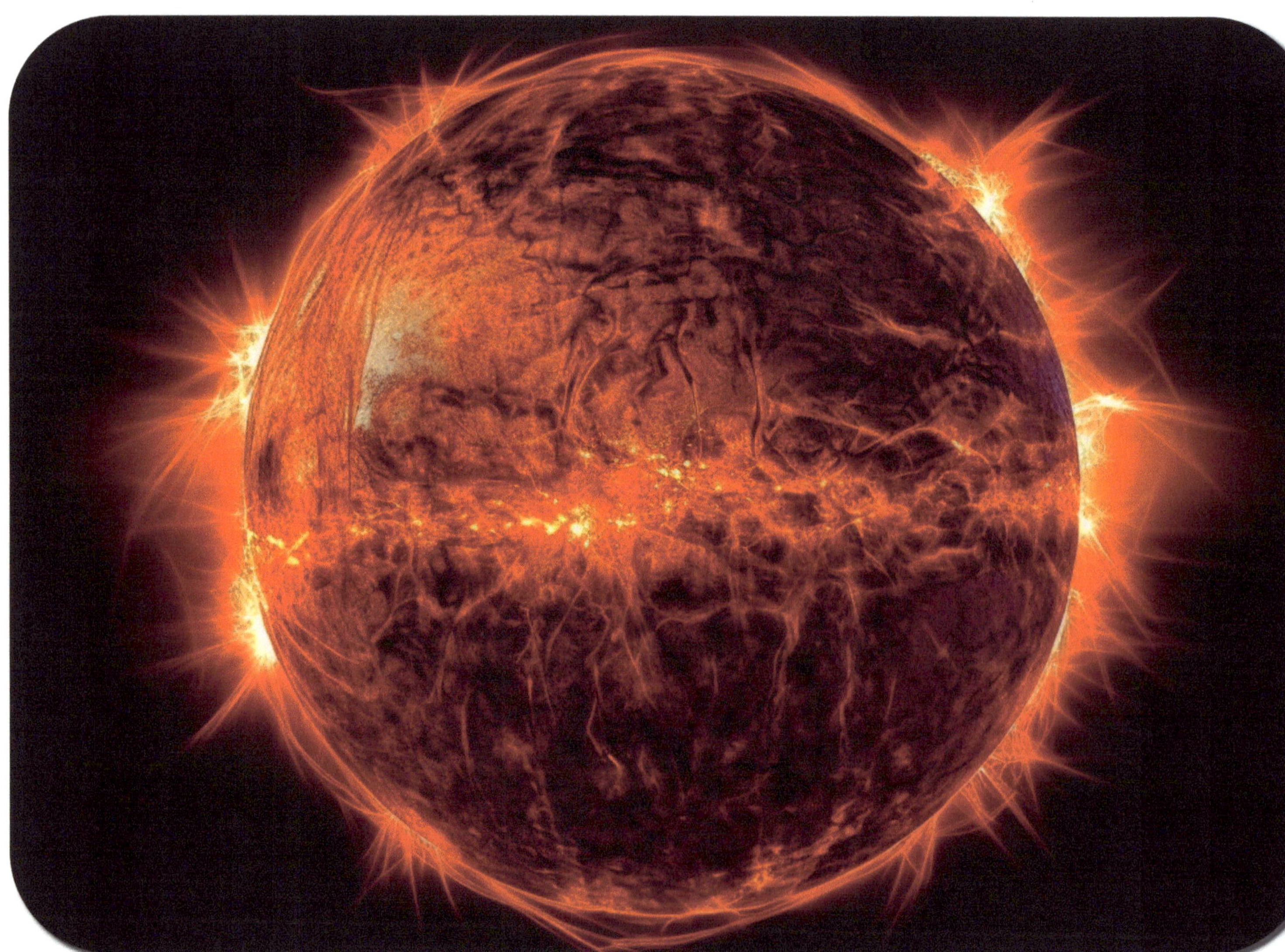

SPACEPOSIUM

The Minor Members:

However, not all celestial objects in a planetary system attain the status of a planet. Dwarf planets like Pluto, Eris, and Haumea, though significant, do not dominate their orbits in the same way the primary planets do. Alongside these, the solar system is teeming with smaller bodies - asteroids in the asteroid belt between Mars and Jupiter, icy comets that visit the inner solar system from the distant Oort Cloud, and the Kuiper Belt objects, remnants from the solar system's formation.

Beyond Our Solar Backyard:

With the dawn of the 21st century and technological leaps in astronomical instruments, our gaze expanded beyond our solar neighborhood. The Kepler Space Telescope and other ground-based observatories spearheaded a revolution in exoplanet discovery. These far-off worlds, ranging from 'Super-Earths' to 'Hot Jupiters,' have showcased our galaxy's astonishing diversity and abundance of planetary systems. With every discovery, we are reminded that our solar system is but one of countless others.

The Quest for Life: The Interplay of Hope, Science, and Silence

Among the most profound questions humanity has ever posed is: "Are we alone?" The discovery of exoplanets, especially those in the 'habitable zone' (a region around a star where conditions may be just right for liquid water), has intensified the search for life. While Earth remains the only known bastion of life, these distant planets offer tantalizing hints that life, in some form or complexity, might be more common than we once thought.

A Universe Teeming with Possibilities

With advancements in technology, especially in the field of astrophysics, we've pinpointed thousands of exoplanets — many nestled in their star's 'habitable zone,' the Goldilocks region where conditions might be conducive for liquid water, a vital ingredient for life as we know it. Given the vast number of stars in our galaxy alone, logic would suggest that life, even in its simplest form, could be widespread.

The Deafening Silence

Yet, herein lies a paradox that has puzzled thinkers, scientists, and enthusiasts alike. Given the apparent high probability of life's existence elsewhere, why have we yet to detect any signs of intelligent civilizations? This puzzle is famously known as the Fermi Paradox, named after the physicist Enrico Fermi. During a conversation about extraterrestrial civilizations, Fermi exclaimed, "Where is everybody?" Indeed, where are the interstellar signals, the spacecraft, or any other tangible evidence of extraterrestrial intelligence?

Potential Explanations: The Fermi Paradox has spurred a multitude of theories. Some postulate that intelligent civilizations tend to self-destruct through technological advancements or other means, while others suggest that they might intentionally avoid or hide from us.

Another perspective is that the vast cosmic distances and the short time-span of technological civilizations might result in our 'missing' each other. The ways we search for extraterrestrial life may not be aligned with how they communicate or exist.

Broader Implications: The search for life and the contemplation of the Fermi Paradox intersect science, philosophy, and spirituality. If we were to find microbial life, our understanding of biology would undergo a paradigm shift. Yet, discovering intelligent extraterrestrial life would profoundly reshape our cultural, religious, and philosophical perspectives, compelling us to view our existence and challenges with a renewed sense of unity.

As we gaze upon the stars, armed with hope and sophisticated tools, we seek answers about extraterrestrial life and ourselves. The quest for life and the enigma of the Fermi Paradox reminds us of the vastness of the unknown, pushing us to explore, learn, and grow. Our journey in this cosmic narrative still unfolds, with each discovery shedding light on yet another piece of the puzzle.

Conclusion: Understanding Our Cosmic Home

As we continue to push the boundaries of science and technology, our understanding of the universe remains in a state of constant evolution. Each advancement in technology and discovery in space exploration expands our knowledge and challenges our preconceptions. The vastness of the cosmos might make us feel insignificant at times, but it also underscores the extraordinary beauty and complexity of our universe.

This ongoing journey is not just about charting the stars or analyzing distant galaxies; it's a profound exploration of our place in the universe. It reminds us that we are part of this grand cosmic tapestry woven together by the very particles and forces we strive to understand. As we look up at the night sky, we are not just spectators but active participants in the unfolding story of the cosmos—a story that inspires us to keep exploring, learning, and wondering.

SPACEPOSIUM

The Infinite Frontier: Our Eternal Quest

As we conclude this chapter, we stand in awe before the unfathomable vastness and complexity of our universe. Its scale defies our everyday comprehension, challenging us to expand our minds and our understanding. In the words of Carl Sagan:

"The significance of a finding that there are other beings who share this universe with us would be absolutely phenomenal. It would be an epochal event in human history."

This quote reminds us of the profound implications our cosmic discoveries can have, not just for science, but for our entire species.

Our journey of discovery has revealed a universe not only vast but also intricately complex. Perhaps most humbling is the realization that despite our remarkable scientific advancements, we still grapple with fundamental questions about the nature of our universal home. Dark matter and dark energy, comprising a staggering 97% of the universe's content, remain enigmatic, reminding us of how much we have yet to learn.

As we gaze into the night sky, we peer not only across vast distances but also back in time. The light from the most distant galaxies we can observe has traveled for over 13 billion years to reach us, offering glimpses of the universe in its infancy. This vast timescale, spanning from the cosmic dawn to the present day, adds yet another dimension to the universe's complexity.

Yet, in the face of this overwhelming vastness and complexity, we find ourselves uniquely positioned. We are the universe becoming aware of itself — conscious observers emerged from the very stardust we study. Our quest to understand the cosmos is not merely an academic pursuit but a profound exploration of our own origins and place in this grand universal design.

In this grand theater of existence, we play the role of both the observer and the observed, the questioner and indeed, it is just beginning. As we look to the future, we are filled with a sense of wonder and anticipation for the mysteries yet to be unraveled.

Let us remember that we are, in the truest sense, children of the stars. Every atom in our bodies was forged in the heart of a star, connecting us intimately to the universe we seek to understand. Our quest is not just about looking outward, but also inward – for in understanding the universe, we come to understand ourselves.

As we close this chapter, let us not see it as an end, but as a beginning – a call to action for future generations to continue this universal odyssey. For in the vast and complex expanse of existence, our story – humanity's story – is still being written, one discovery at a time.

CHAPTER XIII

FUTURE PROSPECTS HUMANITY'S COSMIC DESTINY

As we stand on the precipice of a new era in space exploration, our gaze extends far beyond the familiar confines of our solar system to the vast cosmic ocean that beckons. The future of space exploration is not merely about technological advancement; it is about the expansion of human consciousness, the redefinition of our place in the universe, and the ultimate survival and flourishing of our species.

THE NEXT FRONTIER: CHALLENGES AND OPPORTUNITIES

The immediate future of space exploration presents a myriad of challenges and opportunities that will shape our journey to the stars:

1. Sustainable Space Exploration:

- **Advanced closed-loop life support systems:** These systems aim to recycle and reuse resources like air, water, and waste with minimal loss. They are crucial for long-duration missions where resupply from Earth is impractical or impossible.

- **In-situ resource utilization on other planets**: This involves using local resources on celestial bodies for life support, fuel, and construction materials. It's essential for reducing the mass of supplies that need to be launched from Earth, making extended missions more feasible.

- **Development of space-based solar power:** This technology could provide a constant source of clean energy for both space missions and terrestrial use. Large solar arrays in orbit could collect solar energy and beam it to receivers on Earth or other planets.

2. Artificial Intelligence and Robotics:

- **Autonomous spacecraft for deep space exploration:** These vehicles can make decisions independently, adapting to unexpected situations without real-time input from Earth. This capability is crucial for exploring distant parts of the solar system where communication delays make direct control impractical.

- **AI-driven decision-making systems for space missions:** These systems can process vast amounts of data and make complex decisions faster than human operators. They could optimize resource use, manage spacecraft systems, and respond to emergencies more efficiently than traditional methods.

- **Swarm robotics for planetary exploration and construction:** This involves using large numbers of small, simple robots that work together to accomplish complex tasks. Swarms could explore large areas of a planet's surface or cooperate to build structures, offering flexibility and redundancy impossible with single large robots.

3. Space-Based Manufacturing:

• **Large-scale 3D printing of habitats and spacecraft components:** This technology could allow for the construction of large structures in space without launching them fully assembled from Earth. It offers the potential for more complex and customized designs tailored to the specific needs of space environments.

• **Asteroid mining and processing facilities:** These operations could extract valuable resources from near-Earth asteroids, providing materials for space construction and potentially rare elements for use on Earth. This could significantly reduce the cost of space exploration and open up new economic opportunities.

• **Orbital factories producing unique materials:** The microgravity environment of space allows for the creation of materials with properties not possible on Earth. These could include perfect crystals for electronics, new alloys, or pharmaceuticals with unique structures.

4. Human Adaptation to Space:

• **Genetic engineering for radiation resistance:** This controversial area of research aims to modify human biology to withstand the harsh radiation environment of space better. It could potentially allow for longer missions and reduce long-term health risks for astronauts.

• **Artificial gravity systems for long-duration spaceflight:** These systems, which might involve rotating sections of spacecraft, aim to mitigate the negative health effects of prolonged exposure to microgravity. They could be crucial for maintaining astronaut health on long journeys to Mars or beyond.

• **Brain-computer interfaces for enhanced spacecraft control:** These interfaces could allow astronauts to control spacecraft systems directly with their thoughts. This technology could improve reaction times, allow for multitasking, and provide new ways of interacting with complex systems.

5. Space Traffic Management:

- **AI-powered collision avoidance systems:** As the number of satellites and debris in orbit increases, automated systems for predicting and avoiding collisions become crucial. AI could process vast amounts of tracking data to anticipate and avoid potential collisions more effectively than human operators.
- **International space traffic control networks:** As space becomes more crowded, international cooperation in tracking and managing orbital traffic will be essential. These networks would coordinate global efforts to manage space traffic and prevent collisions.

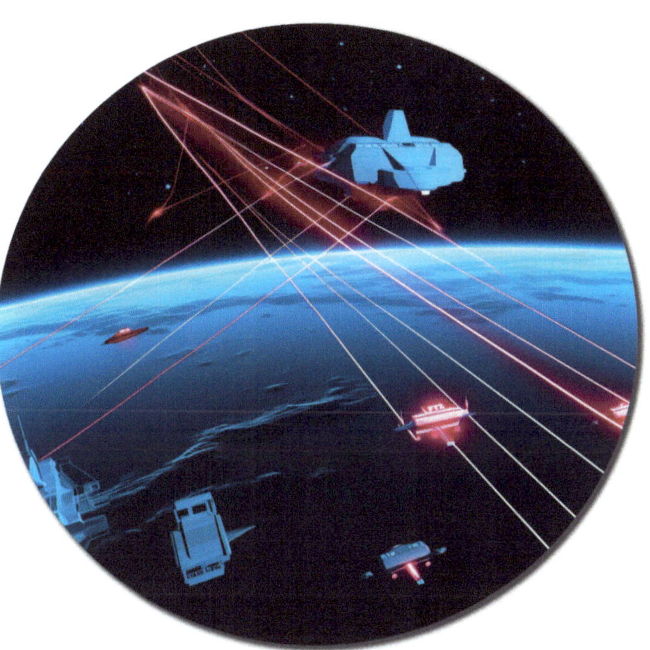

- **Active debris removal technologies:** These systems aim to remove defunct satellites and other debris from orbit. They are crucial for maintaining safe operating conditions in increasingly crowded orbital spaces.

6. Planetary Protection:

- **Advanced sterilization techniques for spacecraft:** These methods aim to prevent contamination of other celestial bodies with Earth microbes. They are crucial for preserving the scientific integrity of exploration and protecting potential extraterrestrial ecosystems.
- **Containment protocols for sample return missions**: These protocols ensure that materials brought back from other celestial bodies do not pose a risk to Earth's biosphere. They involve sophisticated containment and analysis procedures to protect against potential unknown pathogens or harmful substances.
- **Ethical frameworks for interaction with potential extraterrestrial life:** As we explore potentially habitable environments, we need guidelines for how to study and interact with any life we might find. These frameworks would balance scientific curiosity with respect for and protection of alien life forms.

7. Commercial Space Expansion:

- **Orbital hotels and space tourism:** Private companies are developing facilities and vehicles to make space accessible to paying customers. This could range from short suborbital flights to extended stays in orbiting hotels.
- **Private Moon and Martian colonies:** Some companies and organizations plan to establish permanent human settlements on the Moon and Mars. These efforts could pave the way for sustained human presence beyond Earth.
- **Asteroid ownership and resource rights:** As asteroid mining becomes more feasible, legal frameworks for claiming and exploiting space resources will need to be developed. This could lead to a new era of space-based resource extraction and property rights.

INTERSTELLAR DREAMS: THE ULTIMATE FRONTIER

While our solar system offers vast opportunities, humanity's ultimate destiny lies among the stars. Interstellar travel and colonization represent the pinnacle of our species' potential, pushing the boundaries of science, technology, and human endurance:

Propulsion Technologies:

- **Antimatter-catalyzed nuclear pulse propulsion:** This concept uses tiny amounts of antimatter to trigger nuclear fission or fusion reactions, providing immense thrust. It could potentially achieve speeds up to 10% of light speed, making interstellar travel feasible within human lifetimes.
- **Laser-powered light sails:** Enormous ground-based or space-based laser arrays could accelerate ultra-thin sails to relativistic speeds. Projects like Breakthrough Starshot aim to send gram-scale probes to nearby stars within decades.
- **Fusion-driven plasma engines:** Advanced fusion reactors could directly convert fusion energy into thrust, offering high specific impulse and long-duration acceleration. Concepts like the Direct Fusion Drive (DFD) are being developed for potential interplanetary and interstellar missions.

Generation Ships:

- **Self-sustaining arcologies capable of supporting thousands:** These would be massive spacecraft designed as closed ecosystems, recycling all resources and supporting large populations for centuries-long voyages.
- **Artificial ecosystems mimicking Earth's biosphere:** Advanced bioengineering and environmental control systems would recreate Earth-like conditions, including gravity, atmosphere, and diverse ecosystems.
- **Social and governmental systems for multi-generational voyages**: New forms of societal organization would be needed to maintain stability and purpose over multiple generations of space travelers.

Suspended Animation:

- **Cryogenic preservation techniques:** Methods to reversibly freeze and revive human beings for long-duration space travel, potentially using vitrification to avoid cellular damage.
- **Induced hibernation through metabolic manipulation:** Artificially lowering human metabolism to a near-suspended state, similar to how some animals hibernate.
- **Time dilation effects of near-light-speed travel:** At relativistic speeds, time would pass more slowly for the travelers than for those left behind, effectively "stretching" their lifespan relative to Earth time.

Wormhole and Warp Drive Concepts:

- **Manipulation of exotic matter for space-time distortion:** Theoretical concepts that involve using negative energy density to create "shortcuts" through space-time.
- **Alcubierre drive prototypes:** A speculative concept that would contract space in front of a spacecraft and expand it behind, allowing faster-than-light travel without violating relativity.
- **Traversable wormhole stabilization:** Methods to create and maintain stable wormholes as portals between distant points in space.

Interstellar Communication:

- **Quantum entanglement-based instantaneous communication:** Exploiting quantum phenomena to potentially achieve faster-than-light information transfer, though current understanding suggests this may not be possible for transmitting classical information.
- **Neutrino-based signaling for penetrating cosmic distances:** Using neutrinos, which can pass through most matter unimpeded, as a means of long-distance space communication.
- **Gravitational wave modulation for faster-than-light information transfer:** Speculative concepts for using gravitational waves as a communication medium across vast distances.

Exoplanet Exploration and Colonization:

- **Swarm probes for detailed exoplanet surveys:** Large numbers of small, autonomous probes that could comprehensively map and analyze potentially habitable worlds.
- **Terraforming technologies for hostile environments:** Methods to modify entire planetary environments to make them more Earth-like and habitable for humans.
- **Self-replicating von Neumann probes for galactic exploration:** Robotic probes capable of using raw materials to create copies of themselves, potentially exploring the entire galaxy over long timescales.

Ethical and Philosophical Considerations:

- **Protocols for first contact scenarios:** Developing guidelines for how to approach and interact with potential alien civilizations.
- Preservation of human culture and identity across light-years: Ensuring the continuity of human knowledge, values, and cultural diversity over vast distances and timescales.
- **The rights and autonomy of AI and genetically modified colonists:** Addressing the ethical implications of creating new forms of intelligence or significantly altering human biology for space colonization.

These concepts represent the cutting edge of our imagination and scientific understanding, pushing the boundaries of what we believe is possible. While many remain speculative, they serve as guiding stars for future research and development in our quest to become a truly spacefaring civilization.

Conclusion: Our Cosmic Destiny

As we contemplate the future of space exploration, we stand at the threshold of humanity's most incredible adventure. The challenges before us are immense, but they are dwarfed by the limitless potential that awaits us among the stars. From the red sands of Mars to the icy moons of distant gas giants, from the asteroid belts ripe with resources to the tantalizing exoplanets circling alien suns, our destiny is written in the cosmos.

The journey ahead will demand not only technological marvels but also a fundamental reimagining of what it means to be human. As we venture into the depths of space, we will be forced to confront our own limitations, to evolve both physically and mentally, and to forge a new identity as a spacefaring civilization.

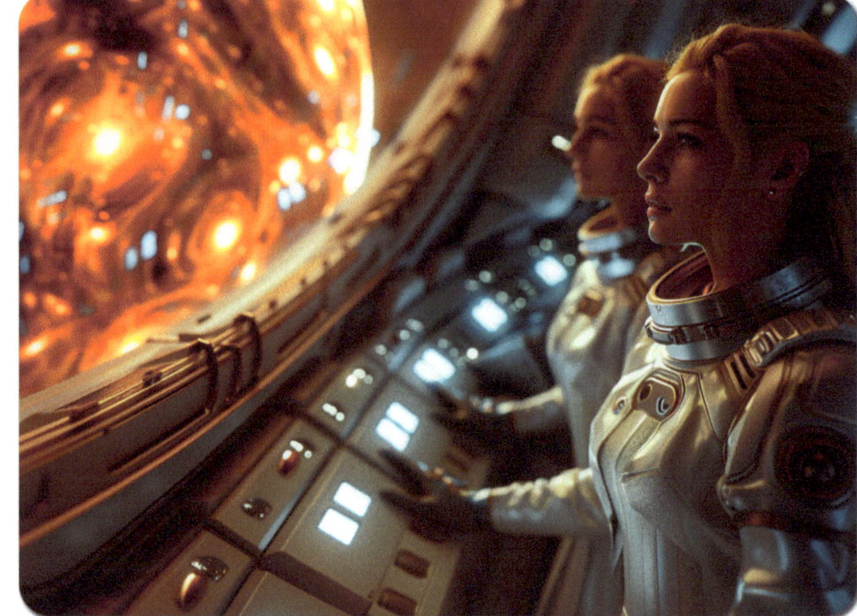

Yet, as we reach for the stars, we must not forget the pale blue dot that gave us birth. Our quest to explore the cosmos must go hand in hand with a renewed commitment to preserve and protect our home planet. The technologies and insights gained from space exploration have the potential to solve many of Earth's most pressing problems, from climate change to resource scarcity.

Stephen Hawking: "For millions of years, mankind lived just like the animals. Then something happened which unleashed the power of our imagination. We learned to talk, and we learned to listen. Speech has allowed the communication of ideas, enabling human beings to work together to build the impossible. Mankind's greatest achievements have come about by talking, and its greatest failures by not talking. It doesn't have to be like this. Our greatest hopes could become reality in the future. With the technology at our disposal, the possibilities are unbounded. All we need to do is make sure we keep talking."

In the words of the visionary Arthur C. Clarke, "The only way of discovering the limits of the possible is to venture a little way past them into the impossible." As we stand on the brink of this new frontier, we are reminded that the greatest discoveries lie not just in the distant reaches of the galaxy but in the boundless potential of the human spirit to imagine, to innovate, and to explore.

The exploration of space is more than a scientific endeavor; it is the ultimate expression of the human spirit. It embodies our insatiable curiosity, our relentless drive to push boundaries, and our capacity to dream of worlds beyond our own. As we set sail on this cosmic ocean, we carry with us the hopes and aspirations of countless generations who have looked up at the night sky in wonder.

In the words of Carl Sagan, "Exploration is in our nature. We began as wanderers, and we are wanderers still. We have lingered long enough on the shores of the cosmic ocean. We are ready at last to set sail for the stars."

The future of space exploration is limited only by our courage to dream and our determination to turn those dreams into reality. As we embark on this grand cosmic odyssey, we do so not just as explorers, but as pioneers of a new chapter in human history. The stars await, and with them, the promise of a future more magnificent than we can possibly imagine.

SPACEPOSIUM

MARS
Mars
Training

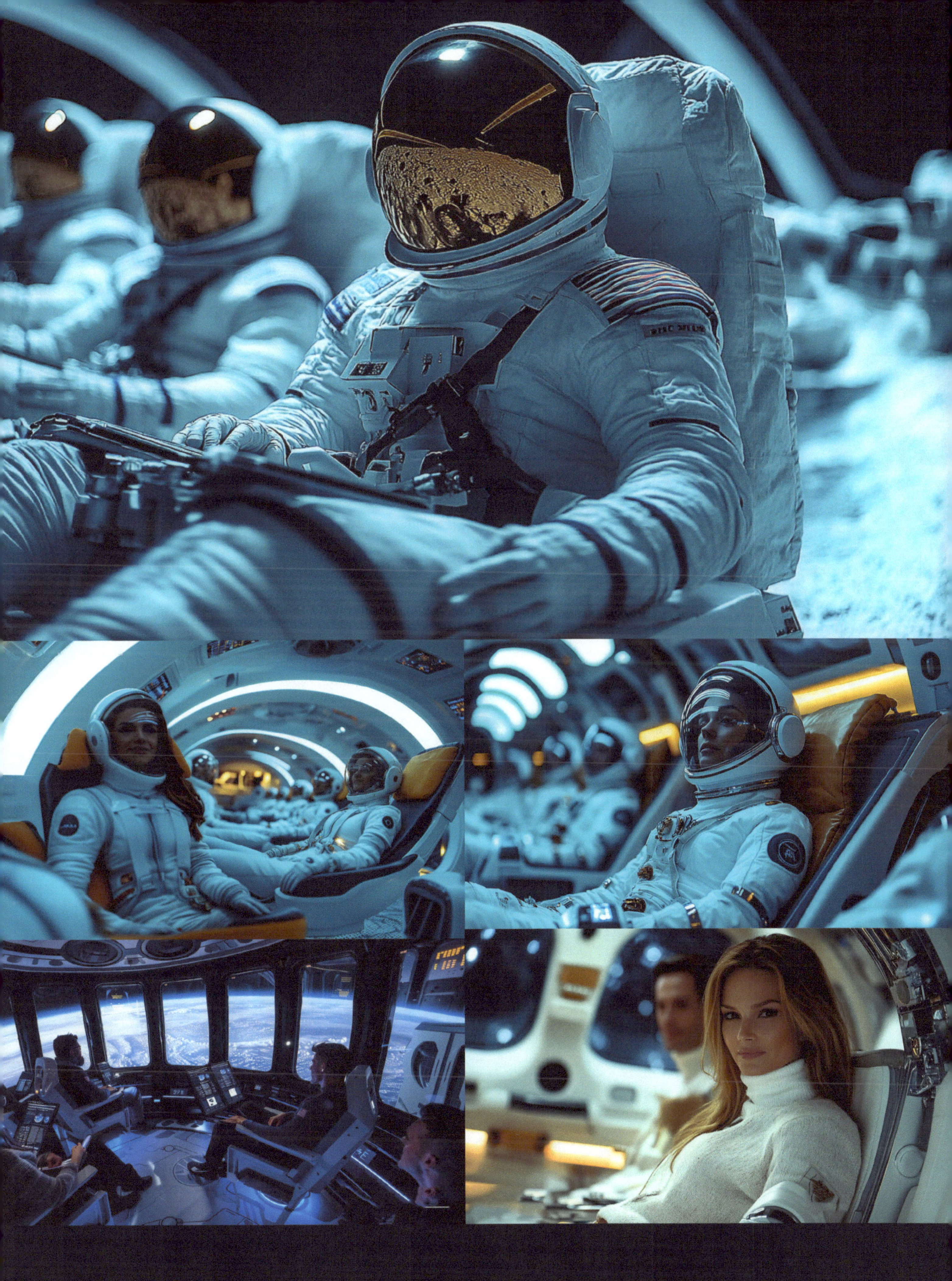

SPACEPOSIUM

MOON
TOURS

SPACE RESOURCES

NAME	DESCRIPTION	URL	Country/City
NASA STEM Engagement	Offers a wide range of STEM resources and activities for K-12 students, educators, and parents.	https://www.nasa.gov/learning-resources/stem-engagement/	USA, Washington, D.C.
NASA Kids' Club	An interactive website with games and activities for young students to learn about space and NASA missions.	https://www.nasa.gov/kidsclub	USA, Washington, D.C.
JPL Education	Provides STEM activities, projects, videos, and games for kids and students.	https://www.jpl.nasa.gov/edu/learn/	USA, Pasadena, CA
JPL Learning Space With NASA at Home	Features educational activities for home-learning and video tutorials.	https://www.jpl.nasa.gov/edu/teach/activity/learning-space-with-nasa-at-home/	USA, Pasadena, CA
SpaceX STEM	Offers educational resources and information about SpaceX missions and technology.	https://www.spacex.com/stem/	USA, Hawthorne, CA
Blue Origin Club for the Future	Provides space-focused activities and educational resources for students and educators.	https://clubforfuture.org/	USA, Kent, WA
Virgin Galactic STEM Resources	Offers educational materials related to space travel and technology.	https://www.virgingalactic.com/learn/	USA, Mojave, CA
European Space Agency (ESA)	Provides educational resources, activities, and competitions for students across Europe.	https://www.esa.int/Education	Europe
Canadian Space Agency (CSA)	Offers educational programs and resources for K-12 students focusing on space science and technology.	https://www.asc-csa.gc.ca/eng/education/default.asp	Canada
JAXA (Japan Aerospace Exploration Agency)	Offers educational resources and programs for students interested in space science and technology.	https://www.jaxa.jp/education/index_e.html	Japan
Indian Space Research Organization (ISRO)	Provides educational resources and outreach programs related to space science in India.	https://www.isro.gov.in/education	India
UK Space Agency	Offers educational resources and activities for students in the UK to engage with space science.	https://www.gov.uk/government/organisations/uk-space-agency	UK
Australian Space Agency	Provides educational resources and programs to inspire students in STEM and space exploration.	https://www.industry.gov.au/strategies-for-the-future/australian-space-agency	Australia
Space Foundation Education	Provides a variety of space-themed educational programs and resources.	https://www.discoverspace.org/education	USA, Colorado Springs, CO
Space Center Houston Education Programs	Offers interactive and hands-on learning environments, including STEM challenges and Innovation Gateway.	https://spacecenter.org/education-programs/	USA, Houston, TX
Space Center Houston Resources	Provides educational activities, Thought Leader Series presentations, and space history resources.	https://spacecenter.org/resources/	USA, Houston, TX
Smithsonian Learning Lab	Extensive collection of digital resources from Smithsonian museums, research centers, libraries, and archives.	https://learninglab.si.edu/	USA, Washington, D.C.
Aldrin Family Foundation	Offers STEAM-based educational tools and programs to inspire the next generation of space explorers.	https://aldrinfoundation.org/	USA, Florida
National Air and Space Museum Educator Resources	Provides curriculum, activities, and resources for educators related to air and space.	https://airandspace.si.edu/learn/educators	USA, Washington, D.C.
FAA Commercial Space Education	Provides resources on commercial space operations, including podcasts and educational materials.	https://www.faa.gov/space	USA, Washington, D.C.
NOAA Data in the Classroom	Offers educational resources using real-time ocean and atmosphere data.	https://dataintheclassroom.noaa.gov/	USA, Silver Spring, MD
Civil Air Patrol Space Lessons	Provides lessons to support US Space Force education.	https://www.gocivilairpatrol.com/	USA, Maxwell AFB, AL

Author Biography

Ben McLaren is a versatile technology and marketing professional turned author, whose diverse career spans technology, marketing, graphic design, audio-visual and AI technologies, and real estate. With a B.A. in Marketing from the University of South Florida and an MBA in Technology Management from the University of Phoenix, Ben's journey is highlighted by a blend of technical acumen and creativity.

At IBM, Ben excelled as an E-Commerce Technical and Sales Specialist, known for transforming proposals into engaging, magazine-like presentations. His career took him to Europe, where he supported IT and educational initiatives for 80 schools across 5 districts, serving over 30,000 students and 6,000 educators.Ben's writing journey began in college and flourished during his time at IBM. His experiences as a photographer in Wellington, FL, and his interactions with successful authors in Palm Beach County inspired him to pursue writing more seriously. Living and working in Germany further fueled his passion, leading to the creation of ImmoAmerica Magazine and his book "International Commissions".

A lifelong science fiction enthusiast, like most people, Ben grew up watching Star Trek, sci-fi movies, and the Apollo missions. His fascination with space continued into adulthood, culminating in a visit to NASA's Johnson Space Center in 2023. A visit to an Open House at NASA's Johnson Space Center in 2023, where Ben witnessed a solar eclipse among thousands of students from all over Texas sparked the inspiration for his latest book, "Spaceposium." This work explores what's going on in Earth's orbit as well as current and future technologies driving the space industry and humanity's potential to live and work beyond Earth. Ben created Spaceposium.com and Spaceposium.org/Spaceposium Foundation, a non-profit organization dedicated to providing educational and inspirational opportunities for students, who aspire to become future space industry professionals. He is now organizing a Global Space Expo for Students in 2025, with an initiative to deploy an audio-visual equipment update to improve the chances for every Student and Educator to enjoy this internationally streamed virtual space expo event.

Ben's global perspective, shaped by living in Germany, California, Italy, and Florida, infuses his writing with a unique blend of cultural insights and technological expertise. His journey from marketing student to multifaceted professional and author reflects his adaptability, creativity, and passion for bridging gaps between industries and cultures.

Contact Information
- **Email**: ben@spasceposium.com
- **Social Media**: https://www.linkedin.com/in/ benmclaren
- https://spaceposium.com | https://spaceposium.org

MARS
The MOON
ROBOTS
The New Face of Automated Industry
A Photographic Odyssey Through the Robot Revolution
MARS
A Visual Odyssey of Innovation and Adventure on Mars
SPACE
The New Age of Space Technology
Ben A McLaren
SPACE
The New Age of Space Technology
Ben McLaren
SPACE
The New Age of Space Technology
Ben McLaren
SPACE
The New Age of Space Technology
Ben McLaren
SPACE
The New Age of Space Technology
Ben A McLaren

Acknowledgment
I would like to express my heartfelt gratitude and appreciation to my friend Katie.
Your belief in this vision has been a driving force behind my determination to create this book.
Thank you for your encouragement, motivation, and insightful feedback.

ENJOY SPACE